SEVEN DAY LOAN

This book is to be returned on
or before the date stamped below

2 9 APR 2003

26/9/03

-8 OCT 2003

14/10/03

13 JAN 2004

29 JAN 2004

1 5 MAR 2004

3 0 MAR 2004

3 0 SEP 2004

3 0 SEP 2004

UNIVERSITY OF PLYMOUTH

PLYMOUTH LIBRARY

Tel: (01752) 232323
This book is subject to recall if required by another reader
Books may be renewed by phone
CHARGES WILL BE MADE FOR OVERDUE BOOKS

SOCIETY FOR EXPERIMENTAL BIOLOGY
SEMINAR SERIES: 62

FISH STRESS AND HEALTH IN AQUACULTURE

SOCIETY FOR EXPERIMENTAL BIOLOGY SEMINAR SERIES

A series of multi-author volumes developed from seminars held by the Society for Experimental Biology. Each volume serves not only as an introductory review of a specific topic, but also introduces the reader to experimental evidence to support the theories and principles discussed, and points the way to new research.

FISH STRESS AND HEALTH IN AQUACULTURE

Edited by

G.K. Iwama
Department of Animal Science, University of British Columbia

A.D. Pickering
Institute of Freshwater Ecology, Cumbria, United Kingdom

J.P. Sumpter
Department of Biology and Biochemistry, Brunel University

C.B. Schreck
Oregon Cooperative Fishery Research Unit, Oregon State University

CAMBRIDGE
UNIVERSITY PRESS

PUBLISHED BY THE PRESS SYNDICATE OF THE UNIVERSITY OF CAMBRIDGE
The Pitt Building, Trumpington Street, Cambridge CB2 1RP

CAMBRIDGE UNIVERSITY PRESS
The Edinburgh Building, Cambridge CB2 2RU, United Kingdom
40 West 20th Street, New York, NY 10011-4211, USA
10 Stamford Road, Oakleigh, Melbourne 3166, Australia

First published 1997

Printed in the United Kingdom at the University Press, Cambridge

Typeset in Linotronic Times 10/12 pt

A catalogue record for this book is available from the British Library

Library of Congress Cataloguing in Publication data

Fish Stress and Health in Aquaculture/edited by G.K. Iwama . . . [et al.].
    p.   cm. – (Society for Experimental Biology seminar series; 62)
    Includes index.
    ISBN 0 521 55518 3 (hc)
    1. Fishes–Effect of stress on.   I. Iwama, G. K. (George K.)
II. Series: Seminar series (Society for Experimental Biology
(Great Britain)); 62.
SH177. S75F57   1997
639.3–dc20   96–44740 CIP

ISBN 0 521 55518 3 hardback

Contents

Measurements of stressed states in the field   247
J.D. MORGAN and G.K. IWAMA

Contributors

BALM, P.H.M.
Department of Animal Physiology, Faculty of Science, University of Nijmegen, Toernooiveld 1, 6525 ED Nijmegen, The Netherlands.

BARTON, B.A.
Department of Biology, University of South Dakota, Vermillion, SD 57069, USA.

DAVIS, M.W.
Cooperative Institute for Marine Resource Studies, Alaska Fisheries Science Center, National Marine Fisheries Service, Hatfield Marine Science Center, Newport, Oregon 97365, USA.

FLETCHER, T.C.
Department of Zoology, University of Aberdeen, Tillydrone Avenue, Aberdeen AB24 2TZ, UK.

IWAMA, G.K.
Department of Animal Science, Canadian Bacterial Diseases Network, University of British Columbia, 208–2357 Main Mall, Vancouver BC, Canada V6T 1Z4.

McDONALD, G.
Department of Biology, McMaster University, 1280 Main St. West, Hamilton, Ontario, Canada L8S 4K1.

MILLIGAN, L.
Department of Zoology, University of Western Ontario, London, Ontario, Canada N6A 5B7.

MORGAN, J.D.
Department of Animal Science, Canadian Bacterial Diseases Network, University of British Columbia, 248–2357 Main Mall, Vancouver, BC, Canada V6T 1Z4.

OLLA, B.L.
Cooperative Institute for Marine Resource Studies, Alaska Fisheries Science Center, National Marine Fisheries Service, Hatfield Marine Science Center, Newport, Oregon 97365, USA.

PANKHURST, N.W.
Department of Aquaculture, University of Tasmania at Launceston, PO Box 1214, Launceston, Tasmania 7250, Australia.
PICKERING, A.D.
The Institute of Freshwater Ecology, Windermere Laboratory, Far Sawrey, Ambleside, Cumbria LA22 0LP, UK.
POTTINGER, T.G.
The Instittue of Freshwater Ecology, Windermere Laboratory, Far Sawrey, Ambleside, Cumbria LA22 0LP, UK.
SCHRECK, C.B.
Oregon Cooperative Fishery Research Unit, U.S. National Biological Service, Department of Fisheries and Wildlife, 104 Nash Hall, Oregon State University, Corvallis, Oregon 97331–3803, USA.
SUMPTER, J.P.
Department of Biology and Biochemistry, Brunel University of West London, Uxbridge, Middlesex, UB8 3PH, UK.
VAN DER KRAAK, G.
Department of Zoology, University of Guelph, Guelph, Ontario, Canada, N1G 2W1.
WEDEMEYER, G.A.
NW Biological Science Center, National Biological Service, 6505 N. E. 65th Street, Seattle, Washington 98115, USA.

Preface

Stressors in intensive aquaculture are unavoidable. While severe stress can result in massive mortalities, sublethal stress can compromise various physiological and behavioural functions, leading to suppressed disease resistance and growth rate, both contributing to suboptimal production. The recognition of stressed states, as well as the management of fish health are therefore critical to the success of an aquaculture operation. The objective of this book is to review current knowledge about stress and health in fish in the context of aquaculture. There is a need to update the knowledge in this field, as it has been more than a decade since a similar volume on fish stress was published. This book also represents some of the papers that were presented as a well-attended Symposium with this theme at the annual meeting of the Society for Experimental Biology in Canterbury. Thus, the book opens with chapters on the physiology of the stress response in fish, and about stress in aquaculture. The genetic basis for the stress response and the effect of nutrition on fish health are reviewed in two chapters. The effects of stress on fish behaviour, immune function and on ionic and osmotic regulation are also presented in respective chapters. Finally, there is a review of current techniques on the detection of stressed states in fish in the field because it is here that the application of knowledge occurs. While the initial focus of this volume was stress in fish husbandry, the chapters are holistic and we believe that others interested in fish biology and management will also find the book of value.

We wish to thank the authors of the individual chapters for their contribution. We believe that their excellent reviews contribute to our understanding of the nature of stress in fish, and to the importance of this knowledge in growing and managing fish through aquaculture as well as other endeavours. We also thank reviewers for their helpful comments on drafts on each chapter. Finally, we wish to thank Paige Ackerman and Grace Cho, who were both critical to this project.

B.A. BARTON

Stress in finfish: past, present and future – a historical perspective

Introduction

The notion that stress is an important consideration in the successful husbandry of finfish is now commonplace among aquaculturists. Indeed, stress and its effects in fish have been the subject of numerous reviews (Mazeaud, Mazeaud & Donaldson, 1977; Pickering, 1981*a*; Adams, 1990*a*; Wedemeyer, Barton & McLeay, 1990; Barton & Iwama, 1991; Fagerlund, McBride & Williams, 1995). The increasing concern about stress shown by practising fish culturists is associated with the advent of modern investigative methods and the application of subsequent research knowledge into present-day finfish culture and management. Culture practices such as handling, sorting, grading, transport and poor water quality impose stress on fish (reviewed by Schreck, 1982; Barton & Iwama, 1991; see also chapter by G.A. Wedemeyer, this volume). In fisheries management, the array of stressors that impinge on fish populations not only includes those encountered in aquaculture, but also various methods of fish capture (Harman, Johnson & Green-wald, 1980; Maule *et al.*, 1988; Hopkins & Cech, 1992; Maule & Mesa, 1994; Mitton & McDonald, 1994), physical trauma (Gadomski, Mesa & Olson, 1994; Sverdrup *et al.*, 1994) and exposure of fish to environmental contaminants (reviewed by Cairns, Hodson & Nraigu, 1984; Adams, 1990*a*; Niimi, 1990; Brown, 1993; Folmar, 1993; Kime, 1995).

Stress affects individual fish and fish populations at all levels of organization ranging from biochemical perturbations to changes in community structure (see Adams, 1990*a*). The detrimental effects of stress may therefore be manifested at the whole-organism level or suborgan-ismally and are considered as direct effects on the organism, whereas indirect effects operating at population or community levels may exert their influence on organism-dependent energy or trophic pathways (Adams, 1990*b*). The latter effects, while of high ecological relevance to natural fish communities, are also the most difficult to define and measure. In finfish aquaculture, however, the important effects of stress are relatively straightforward and, at our present level of understanding, usually involve acute physiological responses to one or more aquaculture-related stressors. Thus, the focus of this chapter is on endocrine, notably corticosteroid, and related metabolic responses of fish to disturbances normally encountered under artificial rearing conditions. This focus also reflects the direction that the bulk of past research on stress in aquaculture has taken.

The purposes of this chapter are to review the concept of stress as it has developed through time, and to examine what has been learned from past research on physiological responses of fish to stress and where those findings may lead to the benefit of aquaculture. A number of topics introduced in this chapter will be discussed in detail in subsequent chapters.

Defining stress

Everybody knows what stress is and nobody knows what it is.

(Selye, 1973)

There are few concepts that have evoked as much discussion and disagreement as that of stress when applied to biological systems.

(Pickering, 1981*b*)

A reliable measurement of stress is critical; however, a reliable, acceptable measurement of stress has not been found, perhaps because the concept is applied to so many different phenomena.

(Moberg, 1985)

I am not certain whether one who undertakes this task (of defining the concept of stress) either has an enormous ego, is unmeasurably stupid, or is totally mad.

(Levine, 1985)

These quotes serve to illustrate the difficulties encountered when trying to provide a universal definition of stress that suits all disciplines. One definition of stress that generally seems to work for providing a framework for research is that stress is 'a state produced by an environmental or other factor which extends the adaptive responses of an animal beyond the normal range or which disturbs the normal functioning to such an extent that, in either case, the chances of survival are significantly reduced' (Brett, 1958). In a physiological context, perhaps a simpler but equally useful definition, which is similar to that proposed by Selye (1973), is that stress is the response of an organism to any demand placed on it such that it causes an extension of a physiological state beyond the normal resting state. Others have defined stress simply as any threat to or disturbance of homeostasis (Munck, Guyre & Holbrook, 1984; Hinkle, 1987).

These latter definitions do not necessarily imply that stress is detrimental to the organism. Indeed, most workers in the field consider the stress response to be a form of adaptive response that promotes the best chance of survival in the face of a noxious or threatening situation. It has evolved as an adaptive response to short-term or acute stressors. Thus, the organism's physiological and behavioural responses are appropriately attuned to cope with the stressful situation confronting it. By doing so, resources (e.g. energy) are diverted from immediately nonessential processes (e.g. growth and reproduction) in order to favour survival in face of the challenge or threat. However, should the organism be exposed to a continuous or chronic stressor from which there is no escape, the adaptive value of the response is compromised and detrimental side-effects may become apparent. Barton and Iwama (1991) considered this aspect as the 'maladaptive' component of the overall stress response.

A strict physiological view is probably not valid, however, as there are behavioural or psychological components to stress that are integral to the manner by which the organism responds (see chapter by C.B. Schreck, B.L. Olla and M.W. Davis, this volume, for an account of the behavioural response to stress). In that context, Schreck (1981) viewed that a characteristic physiological stress response would be evoked if the fish experienced 'fright, discomfort, or pain'. However, some stressors cause specific adverse effects by acting systemically such that the fish may not exhibit the typical physiological stress responses normally measured (for example, cortisol, glucose, ions, lactic acid). The failure of fish to elicit characteristic physiological stress responses after exposure to certain and often lethal chemicals (Grant & Mehrle, 1973; Schreck & Lorz, 1978; Iwama, McGeer & Pawluk, 1989) illustrates the nature of problems encountered when trying to formulate a universal definition of stress.

A useful general definition of stress should suit physiologists, who may be concerned with elevations in plasma hormones or metabolites; toxicologists, who may be most interested in induction of mixed-function oxidase enzymes; or fish culturists, whose main concern may be growth or mortality. However, a definition of stress that fits into every discipline's conceptual framework is not only elusive, it may be impossible. Complete agreement is still wanting with respect to the correct terminology. In spite of sound arguments to apply 'stress' or 'stress factor' to mean the stimulus evoking the response (Pickering, 1981*b*; Wedemeyer *et al.*, 1990), most literature now seems to favour the coined word 'stressor' to mean the stimulus. This agreement is one of convenience to avoid confusion with 'stress', the altered state of the organism, and 'stress response', those physiological or behavioural manifestations we can measure as indicative of a stressed state (Barton & Iwama, 1991). Weiner (1992) adopted a more encompassing view, opting for the term 'stressful experience' in place of 'stress' to clearly indicate that the phenomenon itself is a dynamic process integrating not only perception of the stressor to organize a response, but also the knowledge of prior experience in that process. Regardless of precise terminology, the simple common thread running through these semantic gyrations is one of a biological response to a stimulus at some level of organization.

Evolution of the stress concept

The concept of stress has been with us for a long time. Hippocrates considered that an individual's response to disease had two components: pathos, which was the suffering, and ponos, which was the work exerted by the body to cope with the disease (Reite, 1985). In the English language of the 1600s, stress meant 'hardship, straits, adversity, affliction'. In the 1700s and 1800s, stress first came to mean a force exerted on a person, and later the response or resistance of the person to such forces; this was the meaning adopted by physical and engineering sciences (Reite, 1985). As eloquently discussed by Weiner (1992), the underlying idea for a concept of stress may be attributed to Charles Darwin. Darwin (1859) pointed out in his elaboration of the concept of natural selection that the environment was in constant change, challenging and stressful to the organism. Only those organisms able to cope with such natural challenges were capable of surviving to reproduce. In that sense, such stressful experiences acted as selective pressures.

More than a century ago, Claude Bernard first discussed the importance of constancy in the internal environment, or 'milieu intérieur', for maintaining life, the disruption of which caused sickness (Bernard, 1865). Perhaps one of the earliest modern attempts to articulate the concept of stress in a physiological context was that by Cannon (1935) who described the normal resting state of the organism as 'homeostasis'. Threats to homeostasis involved the autonomic nervous system response to a variety of stimuli, or what Cannon referred to as stress. The first major encompassing theory of stress and its effects on the organism was made by Hans Selye. Selye (1936) noticed a nonspecificity and similarity of responses in his test animals regardless of the disturbance, which he considered as the basis of a general response syndrome exclusive of specific responses to individual stimuli. Such was the nucleus of the General Adaptation Syndrome (GAS) paradigm later put forth (Selye, 1950). The GAS has three stages: first, there is an alarm reaction that is the initial response to the stimulus, followed by a stage of resistance as the organism adjusts or compensates for the disturbance to regain homeostasis. If the organism is unable to cope with the resultant stress, it enters the third stage, i.e. exhaustion, which can lead to the development of a pathological condition or death (Selye, 1973).

The GAS has been in and out of favour with stress physiologists and behaviourists since it was first presented and many have now rejected it, partly because of questions about its supposed nonspecificity. Selye's concept of a nonspecific response to a variety of stimuli was seriously questioned by Mason (1971) who concluded from his experiments that, indeed, neuroendocrine responses could be specific and dependent on how the organism perceives the stressor. Mason (1971) argued that it was not the GAS *per se* that was nonspecific, but rather the behavioural responses to different stressful experiences (e.g. fear, discomfort, pain) that were nonspecific and concluded that the stress concept should be a behavioural one, rather than primarily physiological in nature.

There are other perceived 'flaws' in the GAS concept, the main one being the response of the adrenal system and resultant increase in circulating corticosteroids. Whereas Selye (1950) believed that the role of corticosteroids was to enhance the body's defences against stress, others soon discovered that these hormones were anti-inflammatory and immunosuppressive. Munck *et al.* (1984) argued convincingly that if corticosteroids acted to suppress the immune system responses, which are themselves considered adaptive, how could increases in corticosteroids be considered adaptive? To explain that apparent contradiction, Munck *et al.* (1984) considered that, as an adaptive response, the role of corticosteroids is to protect the body from overshoot of its own defence mechanisms.

Moberg (1985) presented a different approach to understanding stress and the GAS. Simply, the central nervous system perceives a stimulus, the stressor, and organizes a biological defence to it, the stress response. If the organism is unable to cope with the change in biological function, it enters into a prepathological state and eventually develops a pathological condition. Regardless of its pitfalls, the GAS paradigm still provides a useful working conceptual framework for discussing stress in fish resulting from aquacultural practices.

Where we have been – past research related to fish culture

Long before the term stress became commonplace in their working vocabulary, fish culturists were aware of practices resulting in improved health and reduced mortality. Most of the early attention given to ways of reducing stress in aquaculture was in fish transport, which has been carried out for many centuries. For example, it is documented that fish were transported in early Chinese and Roman times (Norris *et al.*, 1960). Perhaps the greatest fish transportation efforts mounted were those in the 1800s from the eastern United States of America to California. Because of the vast distances and time involved, culturists soon realized the value of using ice to slow metabolism, providing adequate aeration and fasting the fish before shipment (Norris *et al.*, 1960). Without being aware of it, they were managing for stress in their fish. For example, Shebley (1927) described the procedure of allowing the fish to rest for some time after arriving by truck or train car before they were distributed; otherwise, they would not 'carry so well'.

Fish culturists have known about the dramatic increase in metabolic rate of handled fish for some time. For example, Haskell (1941) clearly documented a drop in ambient dissolved oxygen levels at four different temperatures in fish transport tanks for the first hour after the fish were loaded into the tank. Haskell (1941) concluded from these observations that the most stressful component of fish transport was the loading process, a fact later confirmed unknowingly by more recent studies (Barton, Peter & Paulencu, 1980; Specker & Schreck, 1980; Maule *et al.*, 1988).

Norris *et al.* (1960) outlined the development of transport methods up to that time, describing ways of controlling temperature and water quality, eliminating waste products, the use of drugs and other additives and various other approaches, all of which were designed to reduce stress and get the fish to their destination in the best condition possible. Norris *et al.* (1960) also discussed the notion of discriminate and indiscriminate stresses, a concept earlier introduced by Brett (1958). Discriminate stresses were those that applied to individuals singly, while indiscriminate stress that caused serious debilitation or death to a few fish during transport would also cause stress to lesser, but possibly serious, degrees in the remainder of the fish.

About 40 years ago, fish culturists were becoming increasingly concerned with factors other than the obvious acute lethal effects of adverse conditions that could result in delayed mortality after transport

(Miller, 1951; Wales, 1954; Horton, 1956). Such concerns led to a series of studies by Edgar Black and colleagues to determine the cause (Black, 1956, 1957*a,b,c*; Black & Barrett, 1957). In these investigations, Black documented the relatively rapid increase and recovery of blood lactic acid in transported salmonid fishes. He compared these changes with similar increases that occurred with both handling practices and strenuous exercise, noting that increased blood lactic acid was associated with fish mortality. Although Black admitted the precise cause of death was unknown, he speculated that the accumulation of lactic acid in the circulation resulting from muscular activity associated with transport would cause severe acid–base imbalances. This hyperactivity-induced upset would reduce the blood's carrying capacity for both oxygen and carbon dioxide, and be the likely cause of death (Black, 1958). This was not a new idea as Huntsman (1938), much earlier, had speculated on a similar mortality factor associated with strenuous muscular activity in commercially captured fishes.

Similar studies were conducted in following years with other species or other stressful aspects of fish transport (Black & Conner, 1964; Caillouet, 1968; Fraser & Beamish, 1969). The notion that lactic acidosis was a direct cause of death was discounted, however, by Wood, Turner and Graham (1983) whose work suggested that intracellular acidosis might have been the actual cause of death in such studies. The impacts of stress on acid–base regulation are fully described in the chapter by D.G. McDonald and C.L. Milligan, this volume. Throughout the 1970s and 1980s, other secondary stress indicators such as changes in blood glucose or electrolytes became more popular than lactic acid measurement for determining stress in fish from aquaculture activities (for example, Chavin & Young, 1970; Miles *et al.*, 1974; Aldrin, Messager & Mevel, 1979; Carmichael *et al.*, 1983; Nikinmaa *et al.*, 1983).

With increased knowledge of steroid hormones in fish and the advent of reliable technology to measure them, plasma cortisol quickly became the stress indicator of choice. Early studies (for example, Leloup-Hatey, 1960; Hane *et al.*, 1966; Fagerlund, 1967; Hill & Fromm, 1968) clearly demonstrated the increase in circulating corticosteroids in fish subjected to handling and exercise. Later studies conducted during the 1970s (for example, Spieler, 1974; Fryer, 1975; Simpson, 1975/76; Strange, Schreck & Golden, 1977; Strange, Schreck & Ewing, 1978) confirmed those findings, thus establishing the importance of the hypothalamic–pituitary–interrenal (HPI) axis in the fish's response to stress. In 1977, Madeline Mazeaud and colleagues published the first major synthesis of this type of work and introduced the idea of primary and secondary effects of stress in fish, primary occurring at the endocrine level and secondary at the level of metabolism and osmoregulation (Mazeaud *et al.*, 1977). Their model implied a causal relationship between features of the two physiological levels of effects, some of which have been clearly established, such as cortisol and immunosuppression, and some not, such as cortisol and glucose elevations. Those distinctions were expanded to encompass tertiary or whole-animal responses to stress, which include performance indicators such as swimming capacity or disease resistance (Wedemeyer & McLeay, 1981).

Where we are – status of current research

Research documenting the responses of cortisol and other hormones, and other secondary and tertiary effects in fish subjected to stressors in aquaculture increased rapidly throughout the 1980s and continues. As a result, characteristic corticosteroid responses to a wide range of aquaculture-related disturbances in many fish species are now well known (reviewed by Barton & Iwama, 1991); more than half of this work has been done with salmonid fishes. Along with corticosteroids, the other important endocrine response is the release of catecholamines (reviewed by Mazeaud & Mazeaud, 1981; Gamperl, Vijayan & Boutilier, 1994). Indeed, neuroendocrine control of these two hormonal axes during stress is likely to be interrelated (Axelrod & Reisine, 1984). Although great emphasis has been placed on plasma cortisol and perhaps less on plasma catecholamines as primary stress indicators, many other hormones have also been shown to respond to stress (Table 1; see also chapter by J.P. Sumpter, this volume). Examples of the major secondary physiological features that have also been used as stress indicators in fish are presented in Table 2. During the last 20 years, a number of generalizations about stress responses in fish have emerged mainly resulting from the examination of corticosteroid responses to acute disturbances.

Magnitude and duration of the response

Fish appear to respond to stress in a manner that reflects both the severity and duration of the stressor (Strange *et al.*, 1978; Barton *et al.*, 1980; Foo & Lam, 1993). In other words, a mild transitory stressor appears to evoke a short-lived response, whereas a severe and continuously applied stressor most likely elicits a more extended response of greater magnitude. For example, a brief handling caused a relatively small rise in plasma cortisol in rainbow trout (*Oncorhynchus mykiss*), but severe confinement accompanied by intense handling elicited much

Table 1. *Examples of hormones other than corticosteroids and catecholamines that have been shown to respond to fish-culture-related stressors*

Hormone	References
Adrenocorticotrophic hormone (ACTH)	Sumpter & Donaldson, 1986; Sumpter, Dye & Benfey, 1986; Pickering *et al.*, 1987; Balm *et al.*, 1994; Pottinger, Balm & Pickering, 1995
β-Endorphin	Sumpter, Pickering & Pottinger, 1985; Pottinger *et al.*, 1995
α-Melanocyte-stimulating hormone	Gilham & Baker, 1985; Sumpter *et al.*, 1985, 1986; Pottinger *et al.*, 1995
Thyroxine	Brown, Fedoruk & Eales, 1978; Leatherland & Cho, 1985; Reddy *et al.*, 1995
Gonadotropin	Pickering *et al.*, 1987; Sumpter *et al.*, 1987
Growth hormone	Pickering *et al.*, 1991; Reddy *et al.*, 1995
Prolactin	Spieler & Meier, 1976; Avella, Schreck & Prunet, 1991; Pottinger, Prunet & Pickering, 1992
Serotonin (5-hydroxytryptamine)	Winberg, Nilsson & Olsén, 1992
Somatolactin	Rand-Weaver, Pottinger & Sumpter, 1993; Kakizawa *et al.*, 1995
Testosterone, 11-ketotestosterone	Pickering *et al.*, 1987; Safford & Thomas, 1987; Melotti *et al.*, 1992; Pankhurst & Dedual, 1994
17β-Oestradiol	Pankhurst & Dedual, 1994

higher levels of cortisol that remained high until the end of the experiment, at which time fish were dying, probably from exhaustion (Barton *et al.*, 1980). Strange *et al.* (1978) also obtained similar results with chinook salmon (*O. tshawytscha*).

Cumulative responses to stress

Responses of fish to stress may be cumulative (Carmichael *et al.*, 1983; Flos *et al.*, 1988; Maule *et al.*, 1988). For example, Barton, Schreck and Sigismondi (1986) subjected three groups of chinook salmon to one, two and three handling stressors, respectively, and found that in each case the peak cortisol responses after the final handling were

Table 2. *Some examples of secondary responses that have been used as indicators of stress in fish (modified from Barton & Iwama, 1991)*

Metabolic	Plasma glucose ˋ
	Plasma lactic acid
	Liver and muscle glycogen
	Liver and muscle adenylate energy charge
	Plasma cholesterol
Haematological	Haematocrit
	Leucocrit
	Erythrocyte (RBC) numbers
	Leucocyte numbers
	Lymphocyte:RBC ratio
	Thrombocyte numbers
	Blood clotting time
	Haemoglobin ˎ
Hydromineral	Plasma chloride
	Plasma sodium ˋ
	Plasma potassium ˋ
	Plasma protein ˋ
	Plasma osmolality
Structural	Interrenal cell size and number
	Interrenal cell nuclear diameter
	Gastric tissue morphology
	Organosomatic indices
	Condition factor

additive. This phenomenon was also demonstrated at the secondary level using plasma glucose (Barton *et al.*, 1986; Mesa, 1994) and at the whole-animal level using stimulus-avoidance or predator-avoidance response times as an indicator (Sigismondi & Weber, 1988; Mesa, 1994). However, see section entitled 'Environmental factors affect stress responses' for a discussion of the effect of habituation on the stress response.

Differences in responses to stress

Different species of fish display different levels of responses to stress (Davis & Parker, 1983, 1986; Sumpter, Dye & Benfey, 1986; Pickering & Pottinger, 1989). Although environmental factors can modify the magnitude of stress responses (see section entitled 'Environmental factors affect stress responses'), differences apparent among species may also be attributed, in part, to variations in stress responses among strains (Barton *et al.*, 1986; Pickering & Pottinger, 1989; Pottinger & Moran, 1993), their hybrids (Williamson and Carmichael, 1986; Noga *et al.*, 1994), discrete stocks (Iwama, McGeer & Bernier, 1992), or between wild and hatchery fish, suggesting a 'domestication' effect. Woodward and Strange (1987) demonstrated that wild rainbow trout, in their response to acute confinement, had not only the highest stress-induced cortisol elevation, but also the largest changes in blood glucose and chloride levels compared to hatchery-stock rainbow trout. Caution should be exercised, however, when comparing differences in the magnitude of stress responses among fish groups, as fish that appear to be 'most stressed' using one indicator (e.g. cortisol) may not necessarily be so using another (e.g. glucose) (Barton & Iwama, 1991).

Within a single strain or population, the stress response has a genetic component (Heath *et al.*, 1993) and some fish may be genetically predisposed to exhibit consistently high or low cortisol responses to stress (Pottinger, Pickering & Hurley, 1992). This trait appears to be at least partly heritable (Fevolden, Refstie & Røed, 1991; Fevolden & Røed, 1993; Pottinger, Moran & Morgan, 1994), albeit heritability estimates made to date are low (Fevolden, Refstie & Gjerde, 1993; Fevolden, Røed & Gjerde, 1994). For further details of the genetic basis of the stress response in fish the reader is referred to the chapter by T.G. Pottinger and A.D. Pickering, this volume.

Ontogeny of the stress response

The stage of development of the fish may also influence the extent to which they respond to stress. For example, cortisol responses 1 h after handling were considerably higher in postsmolt than in presmolt coho salmon (*O. kisutch*), and showed a generally increasing trend over the period of parr–smolt transformation (Barton *et al.*, 1985*a*). Moreover, Maule, Schreck and Kaattari (1987) noted that coho salmon smolts appear to be particularly sensitive to stress at this time, which is an important consideration in salmonid aquaculture as this is the time when salmonid smolts are often physically transferred from the freshwater hatchery to marine net-pens. Additional stress imposed on smolting fish may also impair certain physiological changes occurring at this time in preparation for tolerating salt water (Shrimpton & Randall, 1994).

The ability of fish to respond to stress appears to develop very early in their life history. Pottinger and Mosuwe (1994) determined that the

HPI axis in both rainbow and brown trout (*Salmo trutta*) was responsive to acute stress as early as 5 weeks post-hatch. More recently, Barry *et al.* (1995) tracked resting corticosteroids and the stress response in rainbow trout during early larval development and found that the fish exhibited a characteristic cortisol elevation to an acute stressor by 2 weeks post-hatch, 1 week before exogenous feeding. Those investigators also noted that, although the fish did not respond to stress immediately after hatching, the interrenal tissue at this stage was capable of secreting cortisol upon stimulation by adrenocorticotrophic hormone (ACTH) *in vitro* (Barry, Ochiai & Malison, 1995), suggesting that there may be a brief post-hatch period hyporesponsive to stress at which time the HPI axis is not functioning. These larval studies were made using whole-body extractions to measure corticosteroids. Understandably, blood samples would be extremely difficult to obtain from these fish. Thus, nothing is known about secondary physiological responses to stress during this early life stage.

Once established, the magnitude of the stress response may increase with juvenile development of the fish although the evidence for this is largely circumstantial. For example, 20-g rainbow trout were shown to have greater cortisol and glucose responses to an acute stressor compared with 5-g fish from the same population under the same conditions (Barton, Schreck & Barton, 1987). However, this possible increase in capacity to respond to stress associated with development may not be reflected in adult fish, at least for primary responses. Pottinger *et al.* (1995) recently reported that mature male rainbow trout exhibited lower responses of both cortisol and ACTH to acute stress compared with immature fish, suggesting that the feedback equilibrium 'set point' of the HPI axis may be lowered with the onset of maturity.

Environmental factors affect stress responses

Just about every environmental factor examined has been shown to influence the magnitude of stress responses in fish; some of these known to affect corticosteroid and glucose responses are summarized in Table 3. The practical importance of this fact may be related more to the interpretation of results than to providing new insights about stress responses in fish. These other 'nonstress' factors must be considered not only in experimental design, but also when making comparisons among populations or through time with respect to how fish respond to, or are affected by, stress. The knowledge of certain factors that may modify primary or secondary responses to stress also has practical importance from the standpoint of using these responses as

Table 3. *Examples of environmental factors that have been shown to influence the magnitude of corticosteroid and/or hyperglycaemic responses of fish to stress*

Environmental factor	References
Anaesthetics	Strange & Schreck, 1978; Tomasso, Davis & Parker, 1980; Barton & Peter, 1982; Davis, Parker & Suttle, 1982; Limsuwan *et al.*, 1983; Carmichael *et al.*, 1984*b*; Wedemeyer, Palmisano & Salsbury, 1985; Iwama, McGeer & Pawluk, 1989
Acclimation temperature	Umminger & Gist, 1973; Wydoski, Wedemeyer & Nelson, 1976; Strange Schreck & Golden, 1977; Strange 1980; Carmichael *et al.*, 1984*a;* Davis, Suttle & Parker, 1984; Barton & Schreck, 1987*a*
External salinity	Soivio & Oikari, 1976; Strange & Schreck, 1980; Nikinmaa *et al.*, 1983; Redding & Schreck, 1983; Carmichael *et al.*, 1984*b*; Mazik, Simco & Parker, 1991; Barton & Zitzow, 1995
Nutrition	Barton, Schreck & Fowler, 1988; Vijayan & Moon, 1992; Reddy *et al.*, 1995
Water quality	Carmichael *et al.*, 1984*b*; Barton, Weiner & Schreck, 1985*b*; Pickering & Pottinger 1987*a*
Time of day	Carmichael *et al.*, 1984*a*; Davis, Suttle & Parker, 1984; Barton, Schreck & Sigismondi, 1986
Overhead light	Schreck *et al.*, 1985; Wedemeyer, Palmisano & Salsbury, 1985
Fish density	Specker & Schreck, 1980; Schreck *et al.*, 1985
Background colour	Gilham & Baker, 1985

indicators in attempts to mitigate the effects of stress in aquaculture, for example, the use of salts or anaesthetics to reduce stress during handling or transport (Wedemeyer, 1996).

One factor of interest to aquaculture is that of habituation to stress. Captive fish appear to become conditioned or habituated to repeated disturbances (Barton *et al.*, 1987; Schreck *et al.*, 1995). For example, Barton *et al.* (1987) found that after 10 weeks of a single brief disturbance every day, rainbow trout had a significantly reduced cortisol response to handling than did naïve fish. Earlier work with mammals and birds (Frenkl *et al.*, 1968; Tharp & Buuck, 1974; Rees, Harvey & Phillips, 1983, 1985) suggests that in fish such habituation may be due to adaptation of the HPI axis and, thus, lowered ACTH secretion rather than interrenal exhaustion. Moreover, Redding, Patiño and Schreck (1984) noticed that chronic stress-elevated levels of corticosteroids appeared to stimulate their own clearance rate, and work by Maule and Schreck (1991) with selected tissues suggested that corticosteroids also regulate their own receptors. Both daily acute handling and treatment with cortisol downregulated corticosteroid receptors in coho salmon gill tissue, for example (Shrimpton & Randall, 1994). Habituation to repeated stressful disturbances combined with a feeding reward shows promise as a conditioning protocol for improving survival and performance of fish subjected to subsequent stressors such as transport (Schreck *et al.*, 1995; see the chapter by C.B. Schreck, B.L. Olla and M.W. Davis, this volume, for further discussion of behaviour and performance).

Where we are going – future research directions

In aquaculture, the focus has been on hormonal and other biochemical and physiological responses to acute, usually physical, stressors. Those changes are relatively easy to measure experimentally, tend to be short-lived, but may by themselves, if used simply as indicators, have low ecological relevance (Adams, 1990*b*). Research on stress in fish has reached the point where the important 'So what?' questions about the effects of stress on the performance of fish and health of their populations can now be tackled. Three areas stand out in current stress-related research that have particular relevance to aquaculture; these are the effects of stress on metabolism and bioenergetics, the immune system and reproduction.

Stress and metabolism

There is a metabolic cost associated with stress and the resultant increase in oxygen consumption has been demonstrated to occur in fish subjected to physical disturbances (Saunders, 1963; Smit, 1965; Holeton, 1974; Korovin, Zybin & Legomin, 1982; Barton & Schreck, 1987*b*; Fletcher, 1992; Davis & Schreck, in press). The effects of stress can be viewed in the context of reducing the fish's metabolic scope for activity (Fry, 1947). Presumably, if a portion of the fish's energy budget within its scope for activity is required to cope with stress, then less energy will be available for other performance components (Schreck, 1982; Barton & Iwama, 1991). One such performance component of interest directly related to metabolism and affected by stress is growth (Pickering, 1992). Although the suppression of growth rate in fish stressed by high-density rearing conditions has been demonstrated (Andrews *et al.*, 1971; Refstie & Kittelsen, 1976; Refstie, 1977; Trzebia-towski, Filipiak & Jakubowski, 1981; Vijayan & Leatherland, 1988), it is still not clear that this is a directly stress-mediated effect or one related to water quality, feeding, or social interaction. The chapter by N.W. Pankhurst and G.J. Van Der Kraak, this volume, considers the effects of stress on growth in more detail.

Administration of cortisol to fish suppresses their growth rate (Davis *et al.*, 1985; Barton *et al.*, 1987), and evidence suggests that this steroid may be involved in metabolic processes related to growth and bioenergetics. Indeed, positive correlations between stress-induced increases in plasma cortisol and metabolic rate have been observed (Barton & Schreck, 1987*b*; Davis & Schreck, in press) but the functional significance of this pattern, if any, remains unknown. Cortisol has been ascribed to affect activity of certain metabolic enzymes (Davis *et al.*, 1985; Vijayan, Ballantyne & Leatherland, 1991) and to stimulate production of glucose (Leach & Taylor, 1982; van der Boon, van den Thillart & Addink, 1991), although its role in the latter process still remains unclear (Suarez & Mommsen, 1987; Andersen *et al.*, 1991). Future research is needed to resolve whether elevated cortisol concentrations in fish *under conditions of stress* enhance blood glucose levels, either through gluconeogenesis or by affecting peripheral uptake through action on other metabolic hormones.

Altering nutritional components in diet formulations may show promise for improving the ability of fish to resist stress in aquaculture. Fatty acid enrichment, in particular, has been shown to be useful for enhancing stress resistance in larval marine fishes, for example (Kraul *et al.*, 1993; Ako *et al.*, 1994). Those observations may be related to rapid mobilization of energy reserves to cope with stress. Barton, Schreck and Fowler (1988) noticed that chinook salmon fed the diet with the highest lipid content of three diets tested also showed the greatest glucose response to handling, suggesting that the response reflected a possibly higher capacity to respond to stress rather than simply the fish being 'more stressed'. The chapter by T.C. Fletcher, this volume, examines the links between diet, stress and health.

While studies into the mechanisms of stress responses at the suborganism level are useful and necessary, the effect of stress on metabolic scope requires more thorough exploration as well. Knowing how fish respond to and recover from stress relative to their scope for activity in the context of Fry's paradigm (Fry, 1947; Brett & Groves, 1979) will provide a better understanding of the extent to which performance capacity in fish is limited by stress and will help to bridge the gap from physiological and biochemical responses to whole-animal and population effects.

Stress and immunocompetence

Snieszko (1974) once pointed out that three things were required for an overt disease outbreak: the host, the pathogen and the environment. Clearly, stress is a major environmental component and it is well known that stress increases the susceptibility of fish to disease (Snieszko, 1974; Angelidis, Baudin-Laurencin & Youinou, 1987; Pickering, 1987; Peters *et al.*, 1988; Maule *et al.*, 1989). Furthermore, stress reduces both numbers and antibody-production function of circulating lymphocytes in fish (Ellis, 1981; Miller & Tripp, 1982; Ellsaesser & Clem, 1986; Fries, 1986; Pickering & Pottinger, 1987*b*; Maule *et al.*, 1987, 1989). Not surprisingly, corticosteroids secreted in response to stress appear to be the factor involved in both phenomena (Pickering, 1984; Grimm, 1985; Thomas & Lewis, 1987; Tripp *et al.*, 1987; Pickering & Pottinger, 1989; Pickering, Pottinger & Carragher, 1989).

The stress-activated mechanism of immunosuppression in fish is not completely understood, but appears to be mediated through endocrine pathways (see chapter by P.H.M. Balm, this volume, for further details). Stress-elevated cortisol may inhibit the release of interleukin-like factors involved in the differentiation of lymphocytes from their precursor (Kaattari & Tripp, 1987; Tripp *et al.*, 1987). Moreover, both stress and cortisol have been shown to alter the affinity of corticosteroid receptors in lymphocytes (Maule & Schreck, 1991). Stress has also been shown to reduce lysozyme activity in fish (Möck & Peters, 1990), the humoral enzyme that causes lysis of bacterial cells on contact.

The various mechanisms of stress-induced reduction in immunocompetence is a subject of future research that is likely to prove fruitful and to have direct application to aquaculture. One area of ongoing study that may have future application is in genetic selection of disease-resistant fish based on their corticosteroid stress responses (Fevolden, Refstie & Røed, 1992; Fevolden & Røed, 1993; Fevolden *et al.*, 1993; see also the chapter by T.G. Pottinger and A.D. Pickering, this volume). Another research direction is the development of improved diets and dietary additives that may enhance the fish's immune system

and thereby improve their capacity to cope with aquaculture-related
stressors (reviewed by Blazer, 1992; see also the chapter by T.C.
Fletcher, this volume).

Stress and reproduction

Growing evidence shows that stress in fish can have a profound influence
on their reproductive capacity by affecting circulating levels or altering
seasonal patterns of reproductive hormones (see the chapter by N.W.
Pankhurst and G.J. Van Der Kraak, this volume). While this has been
shown for a number of chemical stressors such as low pH (Freeman
et al., 1983), contaminants (Thomas, 1988, 1989) and industrial effluents
(McMaster *et al.*, 1991; Munkittrick *et al.*, 1992) and was reviewed
recently by Kime (1995), this effect has been also documented for
aquaculture-related stressors such as handling and confinement
(Pickering *et al.*, 1987; Safford & Thomas, 1987; Sumpter *et al.*, 1987;
Melotti *et al.*, 1992; Pankhurst & Dedual, 1994). Again, it appears as
if the effects of stress are mediated through the HPI axis. The suppress-
ive effect of chronically elevated cortisol on plasma testosterone, oestradiol and gonadotropin, and pituitary levels of gonadotropin, as well as
on circulating vitellogenin and ovary weight has been shown in experiments *in vitro* (Carragher & Sumpter, 1990) and using cortisol implants
in fish *in vivo* (Carragher *et al.*, 1989; Pickering, 1989), although
Pankhurst, Van Der Kraak and Peter (1995) recently demonstrated *in
vitro* that this action was not directly on ovarian steroidogenesis for
oestradiol. Cortisol does appear to affect oestradiol-binding sites in the
liver (Pottinger & Pickering, 1990), suggesting a mechanism by which
stress may affect production of vitellogenin.

However, there is also evidence to show that both handling stress
and corticosteroids may stimulate final maturation and ovulation in
certain instances (Billard, Bry & Gillet, 1981; Donaldson & Hunter,
1983; Ayson, 1989). Indeed, the increase in plasma corticosteroids in
upstream-migrating adult Pacific salmon has been well known for years,
but its precise role during this time is still not very clear. Is it solely
a stress-related phenomenon, or is cortisol, either alone or in concert
with other hormones, responsible for stimulating tissue degeneration
during this final act? Stress and its effects on fish reproduction and
egg quality for aquaculture is clearly an area that warrants further
study, particularly for developing proper broodstock management protocols, but also in the context of conservation and enhancement of
wild stocks.

Summary

The state of knowledge about stress in fish culture has come a long way since the days of transporting fingerlings by horse and cart, and particularly so in the past two decades. Science is just scratching the surface of understanding on the entire subject of stress, hormones and performance capacity in fish, particularly in the area of reproduction – a research direction that has importance not only for aquaculture, but also for the well being of natural fish populations subjected to an increasing array of environmental disturbances. In order to know in what direction we are heading in the accumulation and application of knowledge about stress in fish for aquaculture, it is useful to know where we have come from. This chapter was intended to provide a brief overview of past research and where we now appear to be. Where we are going should become evident in the remaining chapters of this book.

References

Adams, S.M., ed. (1990*a*). Biological indicators of stress in fish. *American Fisheries Society Symposium Series*, **8**, 1–191.

Adams, S.M. (1990*b*). Status and use of biological indicators for evaluating the effects of stress in fish. *American Fisheries Society Symposium Series*, **8**, 1–8.

Ako, H., Tamaru, C.S., Bass, P. & Cheng-Sheng, L. (1994). Enhancing the resistance to physical stress in larvae of *Mugil cephalus* by the feeding of enriched *Artemia* nauplii. *Aquaculture*, **122**, 81–90.

Aldrin, J.F., Messager, J.L. & Mevel, M. (1979). Essai sur le stress de transport chez le saumon coho juvenile (*Oncorhynchus kisutch*). *Aquaculture*, **17**, 279–89.

Andersen, D.E., Reid, S.D., Moon, T.W. & Perry, S.F. (1991). Metabolic effects associated with chronically elevated cortisol in rainbow trout (*Oncorhynchus mykiss*). *Canadian Journal of Fisheries and Aquatic Sciences*, **48**, 1811–17.

Andrews, J.W., Knight, L.H., Page, J.W., Matsuda, Y. & Brown, E.E. (1971). Interactions of stocking density and water turnover on growth and food conversion of channel catfish reared in intensively stocked tanks. *Progressive Fish-Culturist*, **33**, 197–203.

Angelidis, P., Baudin-Laurencin, F. & Youinou, P. (1987). Stress in rainbow trout, *Salmo gairdneri*: effects upon phagocyte chemiluminescence, circulating leucocytes and susceptibility to *Aeromonas salmonicida*. *Journal of Fish Biology*, **31** (Supplement A), 113–22.

Avella, M., Schreck, C.B. & Prunet, P. (1991). Plasma prolactin and cortisol concentrations of stressed coho salmon, *Oncorhynchus kisutch*, in fresh water or salt water. *General and Comparative Endocrinology*, **81**, 21–27.

Axelrod, J. & Reisine, T.D. (1984). Stress hormones: their interaction and regulation. *Science*, **224**, 452–59.

Ayson, F.G. (1989). The effect of stress on spawning of brood fish and survival of larvae of the rabbitfish, *Siganus guttatus* (Block). *Aquaculture*, **80**, 241–46.

Balm, P.H., Pepels, P., Helfrich, S., Hovens, M.L.M. & Wendelaar Bonga, S.E. (1994). Adrenocorticotropic hormone in relation to interrenal function during stress in tilapia (*Oreochromis mossambicus*). *General and Comparative Endocrinology*, **96**, 347–60.

Barry, T.P., Malison, J.A., Held, J.A. & Parrish, J.J. (1995). Ontogeny of the cortisol stress response in larval rainbow trout. *General and Comparative Endocrinology*, **97**, 57–65.

Barry, T.P., Ochiai, M. & Malison, J.A. (1995). *In vitro* effects of ACTH on interrenal corticosteroidogenesis during early larval development in rainbow trout. *General and Comparative Endocrinology*, **99**, 382–87.

Barton, B.A. & Iwama, G.K. (1991). Physiological changes in fish from stress in aquaculture with emphasis on the response and effects of corticosteroids. *Annual Review of Fish Diseases*, **1**, 3–26.

Barton, B.A. & Peter, R.E. (1982). Plasma cortisol stress response in fingerling rainbow trout, *Salmo gairdneri* Richardson, to various transport conditions, anaesthesia, and cold shock. *Journal of Fish Biology*, **20**, 39–51.

Barton, B.A. & Schreck, C.B. (1987a). Influence of acclimation temperature on interrenal and carbohydrate stress responses in juvenile chinook salmon (*Oncorhynchus tshawytscha*). *Aquaculture*, **62**, 299–310.

Barton, B.A. & Schreck, C.B. (1987b). Metabolic cost of acute physical stress in juvenile steelhead. *Transactions of the American Fisheries Society*, **116**, 257–63.

Barton, B.A. & Zitzow, R.E. (1995). Physiological responses of juvenile walleyes to handling stress with recovery in saline water. *Progressive Fish-Culturist*, **57**, 267–76.

Barton, B.A., Peter, R.E. & Paulencu, C.R. (1980). Plasma cortisol levels of fingerling rainbow trout (*Salmo gairdneri*) at rest, and subjected to handling, confinement, transport, and stocking. *Canadian Journal of Fisheries and Aquatic Sciences*, **37**, 805–11.

Barton, B.A., Schreck, C.B., Ewing, R.D., Hemmingsen, A.R. & Patiño, R. (1985a). Changes in plasma cortisol during stress and smoltification in coho salmon, *Oncorhynchus kisutch*. *General and Comparative Endocrinology*, **59**, 468–71.

Barton, B.A., Weiner, G.S. & Schreck, C.B. (1985*b*) Effect of prior acid exposure on physiological responses of juvenile rainbow trout (*Salmo gairdneri*) to acute handling stress. *Canadian Journal of Fisheries and Aquatic Sciences*, **42**, 710–17.

Barton, B.A., Schreck, C.B. & Sigismondi, L.A. (1986). Multiple acute disturbances evoke cumulative physiological stress responses in juvenile chinook salmon. *Transactions of the American Fisheries Society*, **115**, 245–51.

Barton, B.A., Schreck, C.B. & Barton, L.D. (1987). Effects of chronic cortisol administration and daily acute stress on growth, physiological conditions, and stress responses in juvenile rainbow trout. *Diseases of Aquatic Organisms*, **2**, 173–85.

Barton, B.A., Schreck, C.B. & Fowler, L.G. (1988). Fasting and diet content affect stress-induced changes in plasma glucose and cortisol in juvenile chinook salmon. *Progressive Fish-Culturist*, **50**, 16–22.

Bernard, C. (1865). *Introduction à l'étude de la médecine expérimentale*. Ballière et Fils, Paris.

Billard, R., Bry, C. & Gillet, C. (1981). Stress, environment and reproduction in teleost fish. In *Stress and Fish*. Pickering, A.D. (ed.), pp. 185–208. Academic Press, London.

Black, E.C. (1956). Appearance of lactic acid in the blood of Kamloops and lake trout following transportation. *Canadian Fish Culturist*, **18**, 3–10.

Black, E.C. (1957*a*). Alterations in the blood level of lactic acid in certain salmonoid fishes following muscular activity. I. Kamloops trout, *Salmo gairdneri*. *Journal of the Fisheries Research Board of Canada*, **14**, 117–34.

Black, E.C. (1957*b*). Alterations in the blood level of lactic acid in certain salmonoid fishes following muscular activity. II. Lake trout, *Salvelinus namaycush*. *Journal of the Fisheries Research Board of Canada*, **14**, 645–49.

Black, E.C. (1957*c*). Alterations in the blood level of lactic acid in certain salmonoid fishes following muscular activity. III. Sockeye salmon, *Oncorhynchus nerka*. *Journal of the Fisheries Research Board of Canada*, **14**, 807–14.

Black, E.C. (1958). Hyperactivity as a lethal factor in fish. *Journal of the Fisheries Research Board of Canada*, **15**, 573–86.

Black, E.C. & Barrett, I. (1957). Increase in levels of lactic acid in the blood of cutthroat and steelhead trout following handling and live transportation. *Canadian Fish Culturist*, **20**, 13–24.

Black, E.C. & Conner, A.R. (1964). Effects of MS 222 on glycogen and lactate levels in rainbow trout (*Salmo gairdneri*). *Journal of the Fisheries Research Board of Canada*, **21**, 1539–42.

Blazer, V.S. (1992). Nutrition and disease resistance in fish. *Annual Review of Fish Diseases*, **2**, 309–23.

Brett, J.R. (1958). Implications and assessment of environmental stress. In *The Investigation of Fish-Power Problems*. Larkin, P.A. (ed.) pp. 69–83. *H.R. MacMillan Lectures in Fisheries*. University of British Columbia, Vancouver.

Brett, J.R. & Groves, T.D.D. (1979). Physiological energetics. In *Fish Physiology*, Volume 8, Hoar, W.S., Randall, D.J. & Brett, J.R. (eds.) pp. 279–352. Academic Press, New York.

Brown, J.A. (1993). Endocrine responses to environmental pollutants. In *Fish Ecophysiology*. Rankin, J.C. & Jensen, F.B. (eds) pp. 276–96. Fish and Fisheries Series 9. Chapman & Hall, London.

Brown, S., Fedoruk, K. & Eales, J.G. (1978). Physical injury due to injection or blood removal causes transitory elevations of plasma thyroxine in rainbow trout, *Salmo gairdneri*. *Canadian Journal of Zoology*, **56**, 1998–2003.

Caillouet, Jr., C.W. (1968). Lactic acidosis in channel catfish. *Journal of the Fisheries Research Board of Canada*, **25**, 15–23.

Cairns, V.W., Hodson, P.V. & Nraigu, J.O., (eds.) (1984). *Contaminant Effects on Fisheries*. John Wiley and Sons, New York.

Cannon, W. (1935). Stresses and strains of homeostasis. *American Journal of the Medical Sciences*, **189**, 1–14.

Carmichael, G.J., Wedemeyer, G.A., McCraren, J.P. & Millard, J.L. (1983). Physiological effects of handling and hauling stress on smallmouth bass. *Progressive Fish-Culturist*, **45**, 110–13.

Carmichael, G.J., Tomasso, J.R., Simco, B.A. & Davis, K.B. (1984*a*). Confinement and water quality-induced stress in largemouth bass. *Transactions of the American Fisheries Society*, **113**, 767–77.

Carmichael, G.J., Tomasso, J.R., Simco, B.A. & Davis, K.B. (1984*b*). Characterization and alleviation of stress associated with hauling largemouth bass. *Transactions of the American Fisheries Society*, **113**, 778–85.

Carragher, J.F. & Sumpter, J.P. (1990). The effect of cortisol on the secretion of sex steroids from cultured ovarian follicles of rainbow trout. *General and Comparative Endocrinology*, **77**, 403–407.

Carragher, J.F., Sumpter, J.P., Pottinger, T.G. & Pickering, A.D. (1989). The deleterious effects of cortisol implantation on reproductive functions in two species of trout, *Salmo trutta* L. and *Salmo gairdneri* Richardson. *General and Comparative Endocrinology*, **76**, 310–21.

Chavin, W. & Young, J.E. (1970). Factors in the determination of normal serum glucose levels of goldfish, *Carassius auratus* L. *Comparative Biochemistry and Physiology*, **33**, 629–53.

Darwin, C.R. (1859). *On the Origin of Species by Means of Natural Selection, or, the Preservation of Favoured Races in the Struggle for Life*. Murray, London.

Davis, K.B. & Parker, N.C. (1983). Plasma corticosteroid and chloride dynamics in rainbow trout, Atlantic salmon, and lake trout during and after stress. *Aquaculture*, **32**, 189–94.

Davis, K.B. & Parker, N.C. (1986). Plasma corticosteroid stress response of fourteen species of warmwater fish to transportation. *Transactions of the American Fisheries Society*, **115**, 495–99.

Davis, K.B., Parker, N.C. & Suttle, M.A. (1982). Plasma corticosteroids and chlorides in striped bass exposed to tricaine methanesulfonate, quinaldine, etomidate, and salt. *Progressive Fish-Culturist*, **44**, 205–207.

Davis, K.B., Suttle, M.A. & Parker, N.C. (1984). Biotic and abiotic influences on corticosteroid hormone rhythms in channel catfish. *Transactions of the American Fisheries Society*, **113**, 414–21.

Davis, K.B., Torrance, P., Parker, N.C. & Suttle, M.A. (1985). Growth, body composition, and hepatic tyrosine aminotransferase activity in cortisol-fed channel catfish, *Ictalurus punctatus*, Rafinesque. *Journal of Fish Biology*, **27**, 177–84.

Davis, L.E. & Schreck, C.B. (1997). The energetic response to handling stress in juvenile coho salmon. *Transactions of the American Fisheries Society*, in press.

Donaldson, E.M. & Hunter, G.A. (1983). Induced final maturation, ovulation, and spermiation in cultured fish. In *Fish Physiology*, Volume 9, part B. Hoar, W.S., Randall, D.J. & Donaldson, E.M. (eds.) pp. 351–403. Academic Press, New York.

Ellis, A.E. (1981). Stress and the modulation of defence mechanisms in fish. In *Stress and Fish*, Pickering, A.D. (ed.) pp. 147–69. Academic Press, London.

Ellsaesser, C.F. & Clem, L.W. (1986). Haematological and immunological changes in channel catfish stressed by handling and transport. *Journal of Fish Biology*, **28**, 511–21.

Fagerlund, U.H.M. (1967). Plasma cortisol concentration in relation to stress in adult sockeye salmon during the freshwater stage of their life cycle. *General and Comparative Endocrinology*, **8**, 197–207.

Fagerlund, U.H.M., McBride, J.R. & Williams, I.V. (1995). Stress and tolerance. In *Physiological Ecology of Pacific Salmon*. Groot, C., Margolis, L. & Clarke, W.C. (eds.) pp. 461–503. University of British Columbia Press, Vancouver.

Fevolden, S.E. & Røed, K.H. (1993). Cortisol and immune characteristics in rainbow trout (*Oncorhynchus mykiss*) selected for high or low tolerance to stress. *Journal of Fish Biology*, **43**, 919–30.

Fevolden, S.E., Refstie, T. & Røed, K.H. (1991). Selection for high and low cortisol stress response in Atlantic salmon (*Salmo salar*) and rainbow trout (*Oncorhynchus mykiss*). *Aquaculture*, **95**, 53–65.

Fevolden, S.E., Refstie, T. & Røed, K.H. (1992). Disease resistance in rainbow trout (*Oncorhynchus mykiss*) selected for stress response. *Aquaculture*, **104**, 19–29.

Fevolden, S.E., Nordmo, R., Refstie, T. & Røed, K.H. (1993). Disease resistance in Atlantic salmon (*Salmo salar*) selected for high or low responses to stress. *Aquaculture*, **109**, 215–24.

Fevolden, S.E., Refstie, T. & Gjerde, B. (1993). Genetic and phenotypic parameters for cortisol and glucose stress response in Atlantic and rainbow trout. *Aquaculture*, **118**, 205–16.

Fevolden, S.E., Røed, K.H. & Gjerde, B. (1994). Genetic components of post-stress cortisol and lysozyme activity in Atlantic salmon; correlations to disease resistance. *Fish and Shellfish Immunology*, **4**, 507–19.

Fletcher, C.R. (1992). Stress and water balance in the plaice *Pleuronectes platessa*. *Journal of Comparative Physiology B*, **162**, 513–19.

Flos, R., Reig, L., Torres, P. & Tort, L. (1988). Primary and secondary stress responses to grading and hauling in rainbow trout, *Salmo gairdneri*. *Aquaculture*, **71**, 99–106.

Folmar, L.C. (1993). Effects of chemical contaminants on blood chemistry of teleost fish: a bibliography and synopsis of selected effects. *Environmental Toxicology and Chemistry*, **12**, 337–75.

Foo, J.T.W. & Lam, T.J. (1993). Serum cortisol response to handling stress and the effect of cortisol implantation on testosterone level in the tilapia, *Oreochromis mossambicus*. *Aquaculture*, **115**, 145–58.

Fraser, J.M. & Beamish, F.W.H. (1969). Blood lactic acid concentrations in brook trout, *Salvelinus fontinalis*, planted by air drop. *Transactions of the American Fisheries Society*, **98**, 263–69.

Freeman, H.C., Sangalang, G.B., Burns, G. & McMenemy, M. (1983). The blood sex hormone levels in sexually mature male Atlantic salmon (*Salmo salar*) in the Westfield River (pH 4.7) and the Medway River (pH 5.6), Nova Scotia. *Science of the Total Environment*, **32**, 87–91.

Frenkl, R., Csalay, L., Csalkvary, G. & Zelles, T. (1968). Effect of muscular exertion on the reaction of the pituitary–adrenocortical axis in trained and untrained rats. *Acta Physiologica Academiae Scientiarum Hungaricae*, **33**, 435–38.

Fries, C.R. (1986). Effects of environmental stressors and immunosuppressants on immunity in *Fundulus heteroclitus*. *American Zoologist*, **26**, 271–82.

Fry, F.E.J. (1947). Effects of the environment on animal activity. *University of Toronto Biology Series* 55, *Publication of the Ontario Fisheries Research Laboratory*, No. 68, pp. 1–62. University of Toronto Press, Toronto.

Fryer, J.N. (1975). Stress and adrenocorticosteroid dynamics in the goldfish, *Carassius auratus*. *Canadian Journal of Zoology*, **53**, 1012–20.

Gadomski, D.M., Mesa, M.G. & Olson, T.M. (1994). Vulnerability to predation and physiological stress responses of experimentally descaled juvenile chinook salmon, *Oncorhynchus tshawytscha*. *Environmental Biology of Fishes*, **39**, 191–99.

Gamperl, A.K., Vijayan, M.M. & Boutilier, R.G. (1994). Experimental control of stress hormone levels in fishes: techniques and applications. *Reviews in Fish Biology and Fisheries*, **4**, 215–55.

Gilham, I.D. & Baker, B.I. (1985). A black background facilitates the response to stress in teleosts. *Journal of Endocrinology*, **105**, 99–105.

Grant, B.F. & Mehrle, P.M. (1973). Endrin toxicosis in rainbow trout (*Salmo gairdneri*). *Journal of the Fisheries Research Board of Canada*, **30**, 31–40.

Grimm, A.S. (1985). Suppression by cortisol of the mitogen-induced proliferation of peripheral blood leucocytes from plaice, *Pleuronectes platessa* L. In *Fish Immunology*. Manning, M.J. & Tatner, M.F. (eds.) pp. 263–71. Academic Press, London.

Hane, S., Robertson, O.H., Wexler, B.C. & Krupp, M.A. (1966). Adrenocortical response to stress and ACTH in Pacific salmon (*Oncorhynchus tshawytscha*) and steelhead trout (*Salmo gairdnerii*) at successive stages in the sexual cycle. *Endocrinology*, **78**, 791–800.

Harman, B.J., Johnson, D.L. & Greenwald, L. (1980). Physiological responses of Lake Erie freshwater drum to capture by commercial shore seine. *Transactions of the American Fisheries Society*, **109**, 544–51.

Haskell, D.C. (1941). An investigation on the use of oxygen in transporting trout. *Transactions of the American Fisheries Society*, **70**, 149–60.

Heath, D.D., Bernier, N.J., Heath, J.W. & Iwama, G.K. (1993). Genetic, environmental, and interaction effects on growth and stress response of chinook salmon (*Oncorhynchus tshawytscha*) fry. *Canadian Journal of Fisheries and Aquatic Sciences*, **50**, 435–42.

Hill, C.W. & Fromm, P.O. (1968). Response of the interrenal gland of rainbow trout (*Salmo gairdneri*) to stress. *General and Comparative Endocrinology*, **11**, 69–77.

Hinkle, Jr., L.E. (1987). Stress and disease: the concept after fifty years. *Social Science and Medicine*, **25**, 561–66.

Holeton, G.F. (1974). Metabolic cold adaption of polar fish: fact or artefact? *Physiological Zoology*, **47**, 137–52.

Hopkins, T.E. & Cech, Jr., J.J. (1992). Physiological effects of capturing striped bass in gill nets and fyke traps. *Transactions of the American Fisheries Society*, **121**, 819–22.

Horton, H.F. (1956). An evaluation of physical and mechanical factors important in reducing delayed mortality of hatchery-reared rainbow trout. *Progressive Fish-Culturist*, **18**, 3–14.

Huntsman, A.G. (1938). Overexertion as cause of death of captured fish. *Science*, **87**, 577–78.

Iwama, G.K., McGeer, J.C. & Pawluk, M.P. (1989). The effects of five fish anaesthetics on acid-base balance, hematocrit, blood gases, cortisol, and adrenaline in rainbow trout. *Canadian Journal of Zoology*, **67**, 2065–73.

Iwama, G.K., McGeer, J.C. & Bernier, N.J. (1992). The effects of stock and rearing history on the stress response in juvenile coho salmon (*Oncorhynchus kisutch*). *ICES Marine Science Symposium*, **194**, 67–83.

Kaattari, S.L. & Tripp, R.A. (1987). Cellular mechanisms of glucocorticoid immunosuppression in salmon. *Journal of Fish Biology*, **31** (Supplement A), 129–32.

Kakizawa, S., Kaneko, T., Hasegawa, S. & Hirano, T. (1995). Effects of feeding, fasting, background adaptation, acute stress, and exhaustive exercise on the plasma somatolactin concentrations in rainbow trout. *General and Comparative Endocrinology*, **98**, 137–46.

Kime, D.E. (1995). The effects of pollution on reproduction in fish. *Reviews in Fish Biology and Fisheries*, **5**, 52–96.

Korovin, V.A., Zybin, A.S. & Legomin, V.B. (1982). Response of juvenile fishes to stress factors associated with transfers during fish farming. *Journal of Ichthyology*, **22**, 98–102.

Kraul, S., Ako, H., Brittain, K., Cantrell, R. & Nagao, T. (1993). Nutritional factors affecting stress resistance in the larval mahimahi, *Coryphaena hippurus*. *Journal of the World Aquaculture Society*, **24**, 186–93.

Leach, G.J. & Taylor, M.H. (1982). The effects of cortisol treatment and protein metabolism in *Fundulus heteroclitus*. *General and Comparative Endocrinology*, **48**, 76–83.

Leatherland, J.F. & Cho, C.Y. (1985). Effect of rearing density on thyroid and interrenal gland activity and plasma and hepatic metabolite levels in rainbow trout, *Salmo gairdneri* Richardson. *Journal of Fish Biology*, **27**, 583–92.

Leloup-Hatey, J. (1960). The influence of enforced exercise on the concentration of plasma corticosteroids in a teleost: the carp (*Cyprinus carpio* L.). *Journal de Physiologie, Paris*, **52**, 145–46.

Levine, S. (1985). A definition of stress? In *Animal Stress*, Moberg, G.P. (ed.), pp. 51–69. American Physiological Society, Bethesda, Maryland.

Limsuwan, C., Limsuwan, T., Grizzle, J.M. & Plumb, J.A. (1983). Stress response and blood characteristics of channel catfish (*Ictalurus punctatus*) after anesthesia with etomidate. *Canadian Journal of Fisheries and Aquatic Sciences*, **40**, 2105–12.

Mason, J.W. (1971). A re-evaluation of the concept of nonspecificity in stress theory. *Journal of Psychiatric Research*, **8**, 323–33.

Maule, A.G. & Mesa, M.G. (1994). Efficacy of electrofishing to assess plasma cortisol concentration in juvenile chinook salmon passing hydroelectric dams on the Columbia River. *North American Journal of Fisheries Management*, **14**, 334–39.

Maule, A.G. & Schreck, C.B. (1991). Stress and cortisol treatment changed affinity and number of glucocorticoid receptors in leukocytes and gill of coho salmon. *General and Comparative Endocrinology*, **84**, 83–93.

Maule, A.G., Schreck, C.B. & Kaattari, S.L. (1987). Changes in the immune system of coho salmon (*Oncorhynchus kisutch*) during the parr-to-smolt transformation and after implantation of cortisol. *Canadian Journal of Fisheries and Aquatic Sciences*, **44**, 161–66.

Maule, A.G., Schreck, C.B., Bradford, C.S. & Barton, B.A. (1988). Physiological effects of collecting and transporting emigrating juvenile chinook salmon past dams on the Columbia River. *Transactions of the American Fisheries Society*, **117**, 245–61.

Maule, A.G., Tripp, R.A., Kaattari, S.L. & Schreck, C.B. (1989). Stress alters immune function and disease resistance in chinook salmon (*Oncorhynchus tshawytscha*). *Journal of Endocrinology*, **120**, 135–42.

Mazeaud, M.M. & Mazeaud, F. (1981). Adrenergic responses to stress in fish. In *Stress and fish*, Pickering, A.D. (ed.) pp. 49–75. Academic Press, London.

Mazeaud, M.M., Mazeaud, F. & Donaldson, E.M. (1977). Primary and secondary effects of stress in fish: some new data with a general review. *Transactions of the American Fisheries Society*, **106**, 201–12.

Mazik, P.M., Simco, B.A. & Parker, N.C. (1991). Influence of water hardness and salts on survival and physiological characteristics of striped bass during and after transport. *Transactions of the American Fisheries Society*, **120**, 121–26.

McMaster, M.E., Van Der Kraak, G.J., Portt, C.B., Munkittrick, K.R., Sibley, P.K., Smith, I.R. & Dixon, D.G. (1991). Changes in hepatic mixed-function oxygenase (MFO) activity, plasma steroid levels and age at maturity of a white sucker (*Catostomus commersoni*) population exposed to bleached kraft pulp mill effluent. *Aquatic Toxicology*, **21**, 199–218.

Melotti, P., Roncarati, A., Garella, E., Carnevali, O., Mosconi, G. & Polzonetti-Magni, A. (1992). Effects of handling and capture stress on plasma glucose, cortisol and androgen levels in brown trout, *Salmo trutta morpha fario*. *Journal of Applied Ichthyology*, **8**, 234–39.

Mesa, M.G. (1994). Effects of multiple acute stressors on the predator avoidance ability and physiology of juvenile chinook salmon. *Transactions of the American Fisheries Society*, **123**, 786–93.

Miles, H.M., Loehner, S.M., Michaud, D.T. & Salivar, S.L. (1974). Physiological responses of hatchery reared muskellunge (*Esox masquinongy*) to handling. *Transactions of the American Fisheries Society*, **103**, 336–42.

Miller, N.W, & Tripp, M.R. (1982). The effect of captivity on the immune response of the killifish, *Fundulus heteroclitus* L. *Journal of Fish Biology*, **20**, 301–308.

Miller, R.B. (1951). Survival of hatchery-reared cutthroat trout in an Alberta stream. *Transactions of the American Fisheries Society*, **81**, 35–42.

Mitton, C.J. & McDonald, D.G. (1994). Consequences of pulsed DC electrofishing and air exposure to rainbow trout (*Oncorhynchus mykiss*). *Canadian Journal of Fisheries and Aquatic Sciences*, **51**, 1791–98.

Moberg, G.P. (1985). Biological response to stress: key to assessment of well-being? In *Animal stress*. Moberg, G.P. (ed.) pp. 27–49. American Physiological Society, Bethesda, Maryland.

Möck, A. & Peters, G. (1990). Lysozyme activity in rainbow trout, *Oncorhynchus mykiss* (Walbaum), stressed by handling, transport, and water pollution. *Journal of Fish Biology*, **37**, 873–85.

Munck, A., Guyre, P.M. & Holbrook, N.J. (1984). Physiological functions of glucocorticoids in stress and their relation to pharmacological actions. *Endocrine Reviews*, **5**, 25–44.

Munkittrick, K.R., McMaster, M.E., Portt, C.B., Van Der Kraak, G.J., Smith, I.R. & Dixon, D.G. (1992). Changes in maturity, plasma sex steroid levels, hepatic mixed-function oxygenase activity, and the presence of external lesions in lake whitefish (*Coregonus clupeaformis*) exposed to bleached kraft mill effluent. *Canadian Journal of Fisheries and Aquatic Sciences*, **49**, 1560–69.

Niimi, A.J. (1990). Review of biochemical methods and other indicators to assess fish health in aquatic ecosystems containing toxic chemicals. *Journal of Great Lakes Research*, **16**, 529–41.

Nikinmaa, M., Soivio, A., Nakari, T. & Lindgren, S. (1983). Hauling stress in brown trout (*Salmo trutta*): physiological responses to transport in fresh water or salt water, and recovery in natural brackish water. *Aquaculture*, **34**, 93–99.

Noga, E.J., Kerby, J.H., King, W., Aucoin, D.P. & Giesbrecht, F. (1994). Quantitative comparison of the stress response of striped bass (*Morone saxatilis*) and hybrid striped bass (*Morone saxatilis* × *Morone chrysops* and *Morone saxatilis* × *Morone americana*). *American Journal of Veterinary Research*, **55**, 405–409.

Norris, K.S., Brocato, F., Calandrino, F. & McFarland, W.N. (1960). A survey of fish transportation methods and equipment. *California Fish and Game*, **46**, 5–33.

Pankhurst, N.W. & Dedual, M. (1994). Effects of capture and recovery on plasma levels of cortisol, lactate and gonadal steroids in a natural population of rainbow trout. *Journal of Fish Biology*, **45**, 1013–25.

Pankhurst, N.W., Van Der Kraak, G. & Peter, R.E. (1995). Evidence that the inhibitory effects of stress on reproduction in teleost fish are not mediated by the action of cortisol on ovarian steroidogenesis. *General and Comparative Endocrinology*, **99**, 249–57.

Peters, G., Faisal, M., Lang, T. & Ahmed, I. (1988). Stress caused by social interaction and its effect on susceptibility to *Aeromonas hydrophila* infection in rainbow trout *Salmo gairdneri*. *Diseases of Aquatic Organisms*, **4**, 83–89.

Pickering, A.D. (ed.) (1981*a*). *Stress and Fish*. Academic Press, London.

Pickering, A.D. (1981*b*). Introduction: the concept of biological stress. In *Stress and Fish*. pp. 1–9. Academic Press, London.

Pickering, A.D. (1984). Cortisol-induced lymphocytopenia in brown trout, *Salmo trutta* L. *General and Comparative Endocrinology*, **53**, 252–59.

Pickering, A.D. (1987). Stress responses and disease resistance in farmed fish. In *Aqua Nor 87, Conference 3: Fish Diseases – a Threat to the International Fish Farming Industry*. pp. 35–49. Norske Fiske-oppdretteres Forening, Trondheim.

Pickering, A.D. (1989). Environmental stress and the survival of brown trout, *Salmo trutta*. *Freshwater Biology*, **21**, 47–55.

Pickering, A.D. (1992). Rainbow trout husbandry: management of the stress response. *Aquaculture*, **100**, 125–39.

Pickering, A.D. & Pottinger, T.G. (1987*a*). Poor water quality suppresses the cortisol response of salmonid fish to handling and confinement. *Journal of Fish Biology*, **30**, 363–74.

Pickering, A.D. & Pottinger, T.G. (1987*b*). Crowding causes prolonged leucopenia in salmonid fish, despite interrenal acclimation. *Journal of Fish Biology*, **32**, 701–12.

Pickering, A.D. & Pottinger, T.G. (1989). Stress responses and disease resistance in salmonid fish: effects of chronic elevation of plasma cortisol. *Fish Physiology and Biochemistry*, **7**, 253–58.

Pickering, A.D., Pottinger, T.G., Carragher, J. & Sumpter, J.P. (1987). The effects of acute and chronic stress on the levels of reproductive hormones in the plasma of mature male brown trout, *Salmo trutta* L. *General and Comparative Endocrinology*, **68**, 249–59.

Pickering, A.D., Pottinger, T.G. & Carragher, J.F. (1989). Differences in the sensitivity of brown trout, *Salmo trutta* L., and rainbow trout, *Salmo gairdneri* Richardson, to physiological doses of cortisol. *Journal of Fish Biology*, **34**, 757–68.

Pickering, A.D., Pottinger, T.G., Sumpter, J.P., Carragher, J.F. & LeBail, P.Y. (1991). Effects of acute and chronic stress on the levels of circulating growth hormone in the rainbow trout *Oncorhynchus mykiss*. *General and Comparative Endocrinology*, **83**, 86–93.

Pottinger, T.G. & Moran, T.A. (1993). Differences in plasma cortisol and cortisone dynamics during stress in two strains of rainbow trout (*Oncorhynchus mykiss*). *Journal of Fish Biology*, **43**, 121–30.

Pottinger, T.G. & Mosuwe, E. (1994). The corticosteroidogenic response of brown and rainbow trout alevins and fry to environmental stress during a 'critical period'. *General and Comparative Endocrinology*, **95**, 350–62.

Pottinger, T.G. & Pickering, A.D. (1990). The effect of cortisol administration on hepatic and plasma estradiol-binding capacity in immature female rainbow trout (*Oncorhynchus mykiss*). *General and Comparative Endocrinology*, **80**, 264–73.

Pottinger, T.G., Pickering, A.D. & Hurley, M.A. (1992). Consistency in the stress response of individuals of two strains of rainbow trout, *Oncorhynchus mykiss*. *Aquaculture*, **103**, 275–89.

Pottinger, T.G., Prunet, P. & Pickering, A.D. (1992). The effects of confinement stress on circulating prolactin levels in rainbow trout (*Oncorhynchus mykiss*) in fresh water. *General and Comparative Endocrinology*, **88**, 454–60.

Pottinger, T.G., Moran, T.A. & Morgan, J.A.W. (1994). Primary and secondary indices of stress in the progeny of rainbow trout (*Oncorhynchus mykiss*) selected for high and low responsiveness to stress. *Journal of Fish Biology*, **44**, 149–63.

Pottinger, T.G., Balm, P.H.M. & Pickering, A.D. (1995). Sexual maturity modifies the responsiveness of the pituitary–interrenal axis to stress in male rainbow trout. *General and Comparative Endocrinology*, **98**, 311–20.

Rand-Weaver, M., Pottinger T.G. & Sumpter, J.P. (1993). Plasma somatolactin concentrations in salmonid fish are elevated by stress. *Journal of Endocrinology*, **138**, 509–15.

Redding, J.M. & Schreck, C.B. (1983). Influence of ambient salinity on osmoregulation and cortisol concentration in yearling coho salmon during stress. *Transactions of the American Fisheries Society*, **112**, 800–807.

Redding, J.M., Patiño, R. & Schreck, C.B. (1984). Clearance of corticosteroids in yearling coho salmon, *Oncorhynchus kisutch*, in fresh water and seawater and after stress. *General and Comparative Endocrinology*, **54**, 433–43.

Reddy, P.K., Vijayan, M.M., Leatherland, J.F. & Moon, T.W. (1995). Does RU486 modify hormonal responses to handling stressor and cortisol treatment in fed and fasted rainbow trout? *Journal of Fish Biology*, **46**, 341–59.

Rees, A., Harvey, S. & Phillips, J.G. (1983). Habituation of the corticosterone response of ducks (*Anas platyrhynchos*) to daily treadmill exercise. *General and Comparative Endocrinology*, **49**, 485–89.

Rees, A., Harvey, S. & Phillips, J.G. (1985). Transitory corticosterone responses of ducks (*Anas platyrhynchos*) to exercise. *General and Comparative Endocrinology*, **59**, 100–104.

Refstie, T. (1977). Effect of density on growth and survival of rainbow trout. *Aquaculture*, **11**, 329–34.

Refstie, T. & Kittelsen, A. (1976). Effect of density on growth and survival of artificially reared Atlantic salmon. *Aquaculture*, **8**, 319–26.

Reite, M. (1985). Implantable biotelemetry and social separation in monkeys. In *Animal Stress*. Moberg, G.P. (ed.) pp. 141–60. American Physiological Society, Bethesda, Maryland.

Safford, S.E. & Thomas, P. (1987). Effects of capture and handling on circulatory levels of gonadal steroids and cortisol in the spotted seatrout, *Cynoscion nebulosus*. In *Proceedings of the Third International Symposium on the Reproductive Physiology of Fish*. Idler, D.W., Crim, L.W. & Walsh, J.M. (eds.) pp. 312. Memorial University of Newfoundland, St. John's.

Saunders, R.L. (1963). Respiration of the Atlantic cod. *Journal of the Fisheries Research Board of Canada*, **20**, 373–86.

Schreck, C.B. (1981). Stress and compensation in teleostean fishes: response to social and physical factors. In *Stress and Fish*. Pickering, A.D. (ed.) pp. 295–321. Academic Press, London.

Schreck, C.B. (1982). Stress and rearing of salmonids. *Aquaculture*, **28**, 241–49.

Schreck, C.B. & Lorz, H.W. (1978). Stress response of coho salmon (*Oncorhynchus kisutch*) elicited by cadmium and copper and potential use of cortisol as an indicator of stress. *Journal of the Fisheries Research Board of Canada*, **35**, 1124–29.

Schreck, C.B., Li, H.W., Maule, A.G., Bradford, C.S., Barton, B.A., Sigismondi, L. & Prete, P.J. (1985). Columbia River salmonid outmigration: McNary Dam passage and enhanced smolt quality. *Completion Report, Contract No. DACW68-84-C-0063*. U.S. Army Corps of Engineers, Portland.

Schreck, C.B., Jonsson, L., Feist, G. & Reno, P. (1995). Conditioning improves performance of juvenile chinook salmon, *Oncorhynchus tshawytscha*, to transportation stress. *Aquaculture*, **135**, 99–110.

Selye, H. (1936). A syndrome produced by diverse nocuous agents. *Nature*, **138**, 32.

Selye, H. (1950). Stress and the general adaptation syndrome. *British Medical Journal*, **1(4667)**, 1383–92.

Selye, H. (1973). The evolution of the stress concept. *American Scientist*, **61**, 692–99.

Shebley, W.H. (1927). History of fish planting in California. *California Fish and Game*, **13**, 163–74.

Shrimpton, J.M. & Randall, D.J. (1994). Downregulation of corticosteroid receptors in gills of coho salmon due to stress and cortisol treatment. *American Journal of Physiology*, **267**, R432–38.

Sigismondi, L.A. & Weber, L.J. (1988). Changes in avoidance response time of juvenile chinook salmon exposed to multiple acute handling stresses. *Transactions of the American Fisheries Society*, **117**, 196–201.

Simpson, T.H. (1975/76). Endocrine aspects of salmonid culture. *Proceedings of the Royal Society of Edinburgh (B)*, **75**, 241–52.

Smit, H. (1965). Some experiments on the oxygen consumption of goldfish (*Carassius auratus* L.) in relation to swimming speed. *Canadian Journal of Zoology*, **43**, 623–33.

Snieszko, S.F. (1974). The effects of environmental stress on outbreaks of infectious diseases of fishes. *Journal of Fish Biology*, **6**, 197–208.

Soivio, A., & Oikari, A. (1976). Haematological effects of stress on a teleost, *Esox lucius* L. *Journal of Fish Biology*, **8**, 397–411.

Specker, J.L. & Schreck, C.B. (1980). Stress responses to transportation and fitness for marine survival in coho salmon (*Oncorhynchus kisutch*) smolts. *Canadian Journal of Fisheries and Aquatic Sciences*, **37**, 765–69.

Spieler, R.E. (1974). Short-term serum cortisol concentrations in goldfish (*Carassius auratus*) subjected to serial sampling and restraint. *Journal of the Fisheries Research Board of Canada*, **31**, 1240–42.

Spieler, R.E. & Meier, A.H. (1976). Short-term serum prolactin concentrations in goldfish (*Carassius auratus*) subjected to serial sampling and restraint. *Journal of the Fisheries Research Board of Canada*, **33**, 183–86.

Strange, R.J. (1980). Acclimation temperature influences cortisol and glucose concentrations in stressed channel catfish. *Transactions of the American Fisheries Society*, **109**, 298–303.

Strange, R.J. & Schreck, C.B. (1978). Anesthetic and handling stress on survival and cortisol concentration in yearling chinook salmon (*Oncorhynchus tshawytscha*). *Journal of the Fisheries Research Board of Canada*, **35**, 345–49.

Strange, R.J. & Schreck, C.B. (1980). Seawater and confinement alters survival and cortisol concentration in juvenile chinook salmon. *Copeia*, **1980**, 351–53.

Strange, R.J., Schreck, C.B. & Golden, J.T. (1977). Corticoid stress responses to handling and temperature in salmonids. *Transactions of the American Fisheries Society*, **106**, 213–17.

Strange, R.J., Schreck, C.B. & Ewing, R.D. (1978). Cortisol concentrations in confined juvenile chinook salmon (*Oncorhynchus tshawytscha*). *Transactions of the American Fisheries Society*, **107**, 812–19.

Suarez, R.K. & Mommsen, T.P. (1987). Gluconeogenesis in teleost fishes. *Canadian Journal of Zoology*, **65**, 1869–82.

Sumpter, J.P. & Donaldson, E.M. (1986). The development and validation of a radioimmunoassay to measure plasma ACTH levels in salmonid fishes. *General and Comparative Endocrinology*, **62**, 367–76.

Sumpter, J.P., Pickering, A.D. & Pottinger, T.G. (1985). Stress-induced elevation of plasma α-MSH and endorphin in brown trout, *Salmo trutta* L. *General and Comparative Endocrinology*, **59**, 257–65.

Sumpter, J.P., Dye, H.M. & Benfey, T.J. (1986). The effects of stress on plasma ACTH, α-MSH, and cortisol levels in salmonid fishes. *General and Comparative Endocrinology*, **62**, 377–85.

Sumpter, J.P., Carragher, J., Pottinger, T.G. & Pickering, A.D. (1987). The interaction of stress and reproduction in trout. In *Proceedings of the Third International Symposium on the Reproductive Physiology of Fish*. Idler, D.W., Crim, L.W. & Walsh, J.M. (eds.) pp. 299–302. Memorial University of Newfoundland, St. John's.

Sverdrup, A., Kjellsby, E., Krüger, P.G., Fløysand, R., Knudsen, F.R., Enger, P.S., Serck-Hanssen, G. & Helle, K.B. (1994). Effects of experimental seismic shock on vasoactivity of arteries, integrity of the vascular endothelium and on primary stress hormones of the Atlantic salmon. *Journal of Fish Biology*, **45**, 973–95.

Tharp, G.D. & Buuck, R.J. (1974). Adrenal adaptation to chronic exercise. *Journal of Applied Physiology*, **37**, 720–22.

Thomas, P. (1988). Reproductive endocrine function in female Atlantic croaker exposed to pollutants. *Marine Environmental Research*, **24**, 179–83.

Thomas, P. (1989). Effects of Aroclor 1254 and cadmium on reproductive endocrine function and ovarian growth in Atlantic croaker. *Marine Environmental Research*, **28**, 499–503.

Thomas, P. & Lewis, D.H. (1987). Effects of cortisol on immunity in red drum, *Sciaenops ocellatus*. *Journal of Fish Biology*, **31** (Supplement A), 123–27.

Tomasso, J.R., Davis, K.B. & Parker, N.C. (1980). Plasma corticosteroid and electrolyte dynamics of hybrid striped bass (white bass × striped bass) during netting and hauling. *Proceedings of the World Mariculture Society*, **11**, 303–10.

Tripp, R.A., Maule, A.G., Schreck, C.B. & Kaattari, S.L. (1987). Cortisol mediated suppression of salmonid lymphocyte responses *in vitro*. *Developmental and Comparative Immunology*, **11**, 565–76.

Trzebiatowski, R., Filipiak, J. & Jakubowski, R. (1981). Effect of stock density on growth and survival of rainbow trout (*Salmo gairdneri* Rich.). *Aquaculture*, **22**, 289–95.

Umminger, B.L. & Gist, D.H. (1973). Effects of thermal acclimation on physiological responses to handling stress, cortisol and aldosterone injections in the goldfish. *Comparative Biochemistry and Physiology*, **44A**, 967–77.

van der Boon, J., van den Thillart, G.E.E.J.M. & Addink, A.D.F. (1991). The effects of cortisol administration on intermediary metabolism in teleost fish. *Comparative Biochemistry and Physiology*, **100A**, 47–53.

Vijayan, M.M. & Leatherland, J.F. (1988). Effect of stocking density on the growth and stress-response in brook charr, *Salvelinus fontinalis*. *Aquaculture*, **75**, 159–70.

Vijayan, M.M. & Moon, T.W. (1992). Acute handling stress alters hepatic glycogen metabolism in food-deprived rainbow trout (*Oncorhynchus mykiss*). *Canadian Journal of Fisheries and Aquatic Sciences*, **49**, 2260–66.

Vijayan, M.M., Ballantyne, J.S. & Leatherland, J.F. (1991). Cortisol-induced changes in some aspects of the intermediary metabolism of *Salvelinus fontinalis*. *General and Comparative Endocrinology*, **82**, 476–86.

Wales, J.H. (1954). Relative survival of hatchery and wild trout. *Progressive Fish-Culturist*, **16**, 125–27.

Wedemeyer, G.A. & McLeay, D.J. (1981). Methods for determining the tolerance of fishes to environmental stressors. In *Stress and Fish*. Pickering, A.D. (ed.) pp. 247–75. Academic Press, London.

Wedemeyer, G.A. (1996). Transportation and handling. In *Principles of Salmonid Culture*. Pennell, W. & Barton, B.A. (eds.) pp. 727–58. Elsevier Science B.V., Amsterdam.

Wedemeyer, G.A., Palmisano, A.N. & Salsbury, L.E. (1985). Development of an effective transport media for juvenile spring chinook salmon to mitigate stress and improve smolt survival during Columbia River fish hauling operations. *Final Report, Contract No. DE-A179–82BP-35460*. Bonneville Power Administration, Portland.

Wedemeyer, G.A., Barton, B.A. & McLeay, D.J. (1990). Stress and acclimation. In *Methods for Fish Biology*. Schreck, C.B. & Moyle, P.B. (eds.) pp. 451–89. American Fisheries Society, Bethesda, Maryland.

Weiner, H. (1992). *Perturbing the Organism: the Biology of Stressful Experience*. University of Chicago Press, Chicago.

Williamson, J.H. & Carmichael, G.J. (1986). Differential response to handling stress by Florida, northern, and hybrid largemouth bass. *Transactions of the American Fisheries Society*, **115**, 756–61.

Winberg, S., Nilsson, G.E. & Olsén, K.H. (1992). The effect of stress and starvation on brain serotonin utilization in Arctic charr (*Salvelinus alpinus*). *Journal of Experimental Biology*, **165**, 229–39.

Wood, C.M., Turner, J.D. & Graham, M.S. (1983). Why do fish die after severe exercise? *Journal of Fish Biology*, **22**, 189–201.

Woodward, C.C. & Strange, R.J. (1987). Physiological stress responses in wild and hatchery-reared rainbow trout. *Transactions of the American Fisheries Society*, **116**, 574–79.

Wydoski, R.S., Wedemeyer, G.A. & Nelson, N.C. (1976). Physiological response to hooking stress in hatchery and wild rainbow trout (*Salmo gairdneri*). *Transactions of the American Fisheries Society*, **105**, 601–6.

G.A. WEDEMEYER

Effects of rearing conditions on the health and physiological quality of fish in intensive culture[1]

The aquatic environment

Normal conditions

The physiological processes of fish are carried out under environmental conditions harsher and more restrictive in many ways than those experienced by terrestrial animals. The concentrations of the gasses in the aquatic 'atmosphere,' for example, are highly variable compared to those in the air. Oxygen depletions are not unusual and at times respiration can be difficult. In addition, water serves not only as the respiratory medium, it also suspends the fish, dilutes their toxic metabolic waste products, and dialyzes materials dissolved in their blood. The energy consumed to regulate the latter process (osmoregulation) can amount to a moderate percentage of the dietary calories.

The aquatic environment is also physically more restrictive than the terrestrial atmosphere experienced by fish biologists. The density of water is more than 800 times greater than that of air and a significant number of calories from the diet must be expended simply pumping the respiratory medium over the gills to extract the oxygen required to support life. By way of comparison, the caloric energy expended to breathe air to extract its oxygen is minimal. The high density of water also requires fish to expend a moderate percentage of their dietary calories to overcome frictional drag while swimming. Since drag increases exponentially with velocity, higher swimming speeds become extremely energy demanding. Finally, life under water subjects fish to challenges from physical factors such as supersaturated gasses and water pressure changes that are usually not even a consideration in the terrestrial world. It is true that fish are evolutionally adapted to these conditions, but this does not imply the absence of energy costs. An

[1]Adapted in part from Wedemeyer, G. (1996a). *The Physiology of Fish in Intensive Culture*. Chapman & Hall, New York.

appreciation of the physiological challenges posed by normal conditions in the aquatic environment is helpful in developing priorities and limits for the additional challenges imposed by conditions in intensive aquaculture systems.

Intensive rearing conditions

The first records of aquaculture date from ancient China and the Roman Empire where fish were apparently produced in ponds under static or low-flow conditions. Under such so-called extensive rearing conditions, the water is required to provide physical living space for the fish, supply oxygen dissolved from the atmosphere, dilute toxic metabolic wastes (excreted or dialyzed from the blood), and serve as the medium in which food organisms are naturally propagated. The necessity for the water to perform all these functions limits the biomass of fish that can be produced. Production can be increased somewhat by supplemental feeding, but the oxygen consumption of the fish and the capacity of the water to dilute and assimilate metabolic wastes will quickly become limiting factors. However, a major increase in fish production can still be obtained if a flow of water is provided through the pond. At the point that fish production becomes dependent on a flow of water, it is usually thought of as intensive rather than as extensive rearing (Westers, 1984).

Intensive aquaculture has many advantages over extensive rearing. The water volume is now required to provide only physical living space for the fish. Its flow through the ponds, raceways, or tanks is used to deliver the required amount of dissolved oxygen (DO) and flush away metabolic wastes. Artificial diets formulated to meet specific nutritional requirements and dispensed under controlled conditions can be used to supply food. The fish biomass that can be achieved thus becomes limited mostly by the food fed and the flow rate of the water rather than by its volume and surface area. For these reasons, the historical trend in aquaculture has been toward more intensive conditions, particularly for salmonid aquaculture in developed countries, but also for warm-water species such as carp and catfish (Tucker & Robinson, 1990). However, intensive culture requires careful management of all aspects of the rearing environment and a detailed understanding of the physiological needs of the fish to prevent the stress and disease problems that would otherwise quickly occur.

Sources of stress in intensive aquaculture

Under the more crowded conditions typical of intensive aquaculture systems, physiological challenges from water chemistry alterations, fish culture procedures, and behavioral interactions between fish are added to the normal demands made by the aquatic environment itself. Within limits, fish can survive such physical, chemical, and behavioral challenges by expending energy (Barton & Iwama, 1991). However, the debilitating effects of many of these stress factors on their physiological condition increase exponentially with only arithmetical changes in the stressor in question – temperature by degrees, for example, or DO by milligrams per liter (Wedemeyer, 1996a). Acutely lethal stress will be rare in well managed aquaculture operations. Chronic stress that eventually becomes manifest as adverse effects on health and physiological condition is the more usual problem. The pathophysiological effects of stress, such as impaired resistance to infectious diseases, can be insidious. Fortunately, many of the potential stress factors in the intensive rearing environment can be managed to improve fish health and physiological condition. Of these stress factors, water quality conditions, fish culture procedures, and biological interactions are probably the most important.

Water quality conditions

Specifying the chemical constituents, temperatures, and dissolved gas concentrations that will provide optimum environmental conditions in intensive culture systems is a complex undertaking. First, most water quality information has been developed to define acute and chronic toxicity levels, not to define the concentrations of water quality constituents that will provide optimum rearing conditions. Thus, considerable information is available on the acute and chronic toxicity of dissolved heavy metals, but very little on the trace metal concentrations required to promote physiological health and disease resistance. For example, fluoride and selenium are potentially toxic but may also improve resistance to bacterial kidney disease (BKD) (Lall et al., 1985). The incidence of BKD may also be inversely related to the dissolved calcium and magnesium concentrations or to ionic composition (Fryer & Lannan, 1993). Second, the effects of water quality conditions on health and physiological condition vary considerably with species, size, age, and previous history of exposure to each dissolved constituent in question. Third, the water quality factors themselves (particularly total hardness, pH, DO, and temperature) can greatly alter the physiological effects of other constituents. For example, concentrations of heavy metals (e.g., zinc, copper) that cause lethal gill damage in soft, acidic water become virtually nontoxic in hard, alkaline water (pH > 7, total hardness >200 mg/l as $CaCO_3$). Similarly, water quality constituents that act to increase or decrease gill ventilation rates also affect toxicity

Table 1. *Water chemistry limits recommended to protect the health of cold- and warm-water fish in intensive culture*[a]

Parameter	Recommended limits
Acidity	pH 6–9
Arsenic	<400 µg/l
Alkalinity	>20 mg/l (as $CaCO_3$)
Aluminum	<0.075 mg/l
Ammonia (un-ionized)	<0.02 mg/l
Cadmium[b]	<0.0005 mg/l in soft water; <0.005 mg/l in hard water
Calcium	>5 mg/l
Carbon dioxide	<5–10 mg/l
Chloride[c]	>4.0 mg/l
Chlorine	<0.003 mg/l
Copper[b]	<0.0006 mg/l in soft water; <0.03 mg/l in hard water
Gas supersaturation	<110% total gas pressure (103% salmonid eggs/fry; 102% lake trout)
Hydrogen sulfide	<0.003 mg/l
Iron	<0.1 mg/l
Lead	<0.02 mg/l
Mercury	<0.0002 mg/l
Nitrate (NO_3^-)	<1.0 mg/l
Nitrite (NO_2^-)	<0.1 mg/l
Oxygen	6 mg/l, cold-water fish; 4 mg/l, warm-water fish
Selenium	<0.01 mg/l
Total dissolved solids	<200 mg/l
Total suspended solids	<80 mg/l
Turbidity (NTU)	<20 NTU over ambient levels
Zinc	<0.005 mg/l

[a]After Wedmeyer (1996a).
[b]To protect smolt development of anadromous salmonids: hard water > 100 mg/l total hardness (as $CaCO_3$), soft water <100 mg/l.
[c]To protect against nitrite toxicity in reuse systems using biofilters for ammonia removal.

because the amount of toxicant to which the gill epithelium is physically exposed is thus increased or decreased. In spite of the complex issues involved, however, a consensus does exist for certain of the water quality conditions necessary to protect the health of fish in intensive culture. A summary of these recommendations is presented in Table 1. More detailed information can be found in Wedemeyer (1996a).

Fish culture procedures

In addition to water quality factors, fish culture procedures such as crowding, handling, and transportation can be important sources of stress in intensive aquaculture. Of these, crowding – approaching or exceeding the density tolerance limits of the species in question – is probably the most common. The term crowding is often loosely used to describe the fish loading density (calculated as the weight of fish per unit of water flow), but it more correctly describes a behavioral requirement for physical space. Crowding should be expressed in terms of the weight density or the weight of fish per unit volume of water. Although the two concepts are related, the fish loading density (weight per unit of flow) is actually a measure of the carrying capacity of the water. In turn, carrying capacity is determined by the oxygen consumption rate of the fish and by physiological tolerance to the ammonia, carbon dioxide, and other toxic metabolic wastes they produce. Thus, the stress of high fish loadings is primarily metabolic while the stress of crowding is 'psychological' and determined by the behavioral requirements of the particular species for physical space.

Because economic considerations usually dictate that maximum use be made of both water and space, both fish density (crowding) and fish loading (carrying capacity of the water) can be significant stress factors in intensive aquaculture systems. However, it is the availability of high-quality water at the flow rates needed to provide oxygen and to dilute metabolic wastes that usually limits fish health rather than the availability of physical space (volume). The requirement for adequate space is usually the second limiting factor. Exceptions can occur due to behavioral factors such as domestication and innate physiological tolerance to crowding stress. Thus, more rainbow trout than either Pacific or Atlantic salmon can normally be reared in a given flow of water even though the oxygen consumption of trout ($mg \cdot kg^{-1} \cdot h^{-1}$) is considerably higher than that of salmon held under similar conditions (Wedemeyer, 1996a).

Fish loadings that will minimize metabolic and respiratory stress can be calculated from tables of the rate of oxygen consumption and ammonia excretion of the species in question (Colt & Orwicz, 1991), or from the feeding rate (Watten, 1992). The latter eliminates the need to consider fish size. In both cases, the physiological criterion is a DO concentration above 6 mg/l and an un-ionized ammonia (NH_3) concentration below 0.02 mg/l at the outflow end of the rearing unit (for salmonids). Carrying capacity can also be determined empirically by progressively reducing the water inflow into a rearing unit containing a known weight of fish until the effluent DO decreases to 6 ppm and the ammonia concentration rises to 0.02 mg/l (for salmonids). However, the cumulative oxygen consumption bioassay developed by Meade (1991) gives more precise results. In this method, the oxygen removal that results in a predetermined maximum acceptable reduction in growth, say 10%, is determined and used to calculate the carrying capacity.

Physiological tolerance to crowding (density) has been a more elusive concept to quantify. Densities that are too conservative waste space while rearing densities that violate behavioral requirements for space will result in debilitated health and physiological condition, reduced food conversion, reduced growth, and ultimately a higher mortality. Changes in blood chemistry usually occur in advance of reduced growth or other whole-animal performance factors. For example, coho salmon parr held under laboratory conditions at densities of 16 kg/m^3 suffer mild hyperglycemia and hypochloremia but the condition becomes severe only in fish held at densities of 100 kg/m^3 and above. In smolting coho, crowding at 16 kg/m^3 causes severe blood chemistry disturbances and higher densities may reduce survival (Wedemeyer, 1996a).

Under production conditions, hatcheries routinely rear salmonids at densities of 60–120 kg/m^3 with no adverse effects on the physiological condition (Westers, 1984). Crowding levels in excess of this may also be satisfactory if the water exchange rate (R) is also very high or if the fish are being held in netpens. In the latter case, the water volume is usually large enough to approximate infinite dilution of metabolic wastes. Thus, rainbow trout held in netpens suspended in circular tanks provided with sufficient flow to maintain loadings at 800 g$\cdot$l$^{-1}\cdot$min^{-1} or less could be reared at densities as high as 267 kg/m^3, which is equivalent to a density index of 11 g$\cdot$l$^{-1}\cdot$cm^{-1}, with no adverse effects on growth, condition factor or other clinical sign of physiological stress (Kebus et $al.$, 1992). In many hatcheries, limited water makes the R factors required to achieve densities such as these too high to be practical.

Using health, growth, and food conversion as the index of pathophysiological effects, density index limits (g$\cdot$l$^{-1}\cdot$cm^{-1} of fish length or ft$^3\cdot$in^{-1} of fish length) have also been established for many fish (Piper et $al.$, 1982). Representative examples for salmonids are: rainbow trout, 3.4 g$\cdot$l$^{-1}\cdot$cm^{-1} (0.5 lb$\cdot$ft$^{-3}\cdot$in^{-1}); coho salmon, 2.7 g$\cdot$l^{-1}l$\cdot$cm^{-1} (0.4 lb$\cdot$ft$^{-3}\cdot$in^{-1}); chinook salmon, 2.0 g$\cdot$l$^{-1}\cdot$cm^{-1} (0.3 lb$\cdot$ft$^{-3}\cdot$in^{-1}). That is, chinook salmon require about 1.7 times as much physical space (volume) as do rainbow trout, and coho salmon require 1.3 times as much. Density tolerance recommendations for Atlantic salmon are

complicated by the fact that this species also has a behavioral requirement for discrete feeding territories. That is, availability of bottom
surface area is important as well as volumetric space. Failure to allow
sufficient areal space (kg fish/m^2 surface) results in fin erosion, poor
growth, and reduced survival. At cold (8–10 °C) water temperatures,
the fin erosion can result in the eventual partial or complete loss of
the dorsal, pectoral, and pelvic fins. In warmer water, the damaged
fin tissues can regenerate and higher areal densities can be tolerated.
At the 17–18 °C temperatures commonly used by hatcheries that can
heat water to accelerate growth, juvenile Atlantic salmon can be reared
to a final fish density of up to 21 kg/m^2 (146 kg/m^3) without adverse
effects on growth, survival, or fin condition. In the colder water typical
of natural rearing conditions (8–10 °C), densities should be kept below
21 kg/m^3 (Soderberg, Meade & Redell, 1993). Juvenile Atlantic salmon
reared in fresh water at densities of 8, 32, and 64 kg/m^3 suffered chronic
hypercholesterolemia at 64 kg/m^3 and significantly lower numbers of
antibody-producing (plaque-forming) white blood cells (Mazur &
Iwama, 1993). Such physiological effects may help explain why both
Pacific and Atlantic salmon smolts reared at high densities often suffer
reduced ocean survival after release.

Guidelines for the densities (lb/ft^3, kg/m^3) and flow loadings (lb/gpm
flow, kg·l^{-1}·min^{-1}) that will minimize stress and protect the health and
physiological condition of Pacific anadromous salmonids can be found
in Banks, Taylor and Leek (1979). These loadings were developed to
minimize crowding stress by providing sufficient physical space to maintain the density factor at about 0.35 or below and minimize respiratory
stress by providing sufficient water flow to maintain the DO above
8 mg/l and un-ionized ammonia concentrations below 0.01 mg/l. Other
salmonids, such as rainbow trout, lake trout, and brown trout, are
more resistant to crowding stress and can tolerate density factors up
to 0.5 without compromising disease resistance. However, flow rates
must still be adjusted to maintain the DO above 6 mg/l and un-ionized
ammonia below 0.02 mg/l to minimize respiratory stress. Density guidelines for cool-water and warm-water species such as northern pike,
walleye, channel catfish, and striped bass can be found in Wedemeyer
(1996*a*).

In warm-water fish culture, crowding stress is usually a less important
concern because, other than the channel catfish, these species are
normally reared extensively. In the intensive culture of channel catfish,
densities of up to 48 fish·kg^{-1}·m^{-3} have been successfully achieved
(Collins, Burton & Schweinforth, 1987). Carrying capacities for the
water used in intensive catfish culture are determined using the concepts

already discussed for cold-water fish. Again, the pathogen load of the water supply is highly relevant to both density tolerance and carrying capacity but is rarely used directly because it is so difficult to quantify.

Another fish culture procedure that is an important source of stress is transportation. Economic necessity normally dictates that the smallest possible volume of water be used to transport the largest possible number of fish and correctly designed life support systems must be used to prevent the adverse water quality changes that would otherwise quickly occur.

The most basic physiological requirement is the continuous circulation of freshly aerated water to all parts of the hauling tank. This is absolutely essential to fish health and a water recirculation rate equivalent to at least 0.5 exchanges per minute is widely considered desirable for both cold-water and warm water fish (Wedemeyer, 1996b). An aeration system that continuously replaces the DO consumed by the fish is also an absolute requirement. Salmonid species of importance in aquaculture consume oxygen at $200–400$ $mg \cdot kg^{-1} \cdot h^{-1}$ even under normal conditions and this rate can double if the fish are excited, stressed, or are swimming rapidly. An aeration system that will maintain a DO concentration of at least 80% of saturation is recommended to minimize respiratory stress. Oxygen aeration is normally required to provide the DO needed and to increase the efficiency of stripping carbon dioxide and other unwanted dissolved gases (such as nitrogen) from the hauling tank water. If carbon rod diffusers are used, the oxygen flow rate needed is about 3 liters oxygen per minute per 250 kg fish per meter diffuser length (Wedemeyer, 1996b). A potentially stressful aspect of oxygen aeration is that DO concentrations as high as 16–18 mg/l can quickly occur. This quickly leads to elevated arterial carbon dioxide levels in the transported fish because the high DO reduces the normal gill ventilation rate while increasing the metabolic production of carbon dioxide. Over a period of several hours, the plasma bicarbonate concentration rises correspondingly to compensate for the hypercapnia (Hobe, Wood & Wheatly, 1984). After the fish are released from the hauling truck, the elevated blood gases rapidly diffuse into the receiving water driven by the steep concentration gradient. The blood pH then rises sharply and the resulting alkalosis requires more hours for compensation. Smolts transported in fresh hyperoxic water and released directly into marine netpens suffer blood oxygen changes that are even larger because of the lower DO of sea water. Although these physiological challenges are normally within the tolerance limits of fish in good health, it is prudent to allow a recovery period of several days before subjecting them to additional stress from other fish culture procedures (Pennell, 1991).

The accumulation of ammonia and carbon dioxide in fish hauling tanks can be another source of stress during fish transport operations. Cold-water and warm-water species important to aquaculture, such as salmonids and channel catfish, excrete about 30 g of ammonia and 1.4 mg carbon dioxide for each kilogram of food and milligram of oxygen consumed.

High carbon dioxide concentrations in fish distribution tanks present more serious problems than the accumulation of ammonia. In low alkalinity waters, the concentration of toxic ammonia in the hauling tank will usually remain below the accepted 4-h acute exposure limit of 0.1 mg/l even if the total ammonia concentration reaches 10 mg/l or more because the carbon dioxide that is also continuously being produced by the fish will usually maintain the pH at 7 or below. For example, rainbow trout transported at the normal loading density of 0.3 kg/l (3 lb/gal), can easily produce enough carbon dioxide to raise the initial concentration to 30 mg/l or more within 30 min unless steps are taken to prevent it (Smith, 1978). Failure to remove excess dissolved carbon dioxide results first in hypercapnia and acidosis, then respiratory stress from the Bohr and Root effects, tissue hypoxia, and eventually carbon dioxide narcosis and death (Wedemeyer, 1996b). Concentrations greater than 15–20 mg dissolved carbon dioxide/l are physiologically undesirable during hauling because oxygen transport to the tissues begins to be impaired due to the Bohr and Root effects. Hypercapnia may also overwhelm the blood buffering system resulting in (respiratory) acidosis and further stress. The classic work of Basu (1959) showed that for salmonids, the DO required to provide enough oxygen to the tissues to support a moderate swimming activity level increases from only 6 mg/l if little or no carbon dioxide is present, to more than 11 mg/l at a dissolved carbon dioxide concentration of 30 mg/l. Thus, the usual stress mitigation guideline that (salmonid) fish in distribution units will have adequate oxygen as long as the DO level does not fall below about 80% of saturation is true only if the dissolved carbon dioxide levels are kept below 20–30 mg/l.

Increased blood lactic acid concentrations (hyperlacticemia) due to excessive swimming activity can also cause (metabolic) acidosis and reduced oxygen transport to the tissues due to the Bohr and Root effects. Millimolar quantities of sodium bicarbonate or sodium sulfate added to the hauling tank water will offset the decrease in blood buffering capacity due to the acidosis and help maintain a stable blood pH (Haswell *et al.*, 1982).

Respiratory stress from the Bohr and Root effects can be minimized by fish handling protocols that will prevent hyperlacticemia (by minimizing excitement and swimming activity), and by an aeration system that will prevent hypercapnia by stripping out dissolved carbon dioxide as well as providing an adequate supply of DO. In practice, these are two of the most important considerations in meeting the physiological needs of transported fish.

Of paramount importance in minimizing the stress of fish transport operations is a correct choice of fish loading density (fish weight per unit volume of water). This must usually be determined empirically because it depends on the physiological requirements of the fish to be hauled, the efficiency of the aeration system, the water chemistry, and the hauling tank design. At present, loading density is commonly expressed in at least three different ways. Percent loading (weight of the fish as a percentage of the weight of the water), weight/volume loading (weight of fish per unit volume of water), or the displaced weight loading (weight of fish per unit volume of water minus the volume of water displaced by the fish) are all used regularly. In calculating loading densities by the latter method, the volume of the water displaced is subtracted from the final volume since it is not available to the fish.

For nonanadromous rainbow trout, or brook, brown, and lake trout, hauling densities of about 36% (0.36 kg/l, 3.4 lb/gal) can be safely achieved when transporting larger (25 cm) fish (Piper *et al.*, 1982; Westers 1984). For anadromous salmonid juveniles, carrying capacities in the range of 6–24% (0.06–0.24 kg/l, 0.51–2.0 lb/gal) for fingerlings, and 24–36% (0.24–0.36 kg/l, 2.0–3.0 lb/gal) for the larger parr (10–12 cm) are considered minimally stressful. In smolt hauling, much lower loading densities must be used because anadromous salmonids during this life stage are more sensitive to handling and crowding stress and scale loss (Barton & Iwama, 1991). In hauling smolts by truck, the (conservative) guideline widely used in the United States of America is a weight loading of 5% (0.05 kg/l, 0.4 lb/gal). In barges or ships using pumped river or ocean water, the same weight loading with a flow rate of about 2 $l \cdot min^{-1} \cdot kg^{-1}$ fish (0.24 $gal \cdot min^{-1} \cdot lb^{-1}$) is used to achieve an adequate water exchange rate (Ceballos, Petit & Mckern, 1991).

A variety of stress mitigation techniques have also been developed to minimize the adverse physiological effects of transportation on fish health. These include water additives (anesthetics, mineral salts, and polymers), fasting, hypothermia, and reduced light intensity.

Anesthetic and hypnotic drugs added to the hauling tank water can be of significant help in mitigating physiological stress. Drug additives are primarily used to slow metabolic rates and thus reduce oxygen consumption and ammonia and carbon dioxide production. However, they also mitigate the stress response caused by excitement and handling, and reduce swimming activity as well. The latter fact is a significant help in preventing injuries such as broken fins and scale loss that would ordinarily occur due to hyperactivity. If drugs are to be used, only compounds licensed for fisheries applications should be selected, and proper dosages must be used. For anesthetics and hypnotics, the proper dosage is one that produces light sedation (reduced reaction to external stimuli without loss of equilibrium) rather than anesthesia (loss of sensation and, usually, equilibrium). It is critical that equilibrium and essential physiological functions such as osmoregulation and gill respiratory exchange are not affected (Summerfelt & Smith, 1990). Anesthetized fish which lose equilibrium will sink to the bottom of the hauling tank and may suffocate or be drawn against the pump screens causing scale loss and preventing adequate water circulation. Of the many drugs tested as transport additives, the hypnotics quinalbarbitone (Seconal Sodium) amylobarbitone (Sodium Amytal), quinaldine, etomidate, metomidate, and the anesthetics 2-phenoxyethanol (2-PE) and tricaine methanesulfonate (MS-222) have received the widest use. Using such drugs, rainbow trout have been successfully transported at two to three times the normal weight of fish per unit volume of water. Unfortunately, many drugs are not cleared for use and others have adverse side-effects such as bradycardia and hypertension. Although use of anesthetic or hypnotic drugs can make it possible to increase the fish loading density by a factor of two or three times the normal weight per unit volume of water, these agents should not be used solely for this purpose or to compensate for basic deficiencies in water quality. As general precautions, a recovery period of several days should be allowed before any drug is used a second time for transported fish, and human skin contact should be minimized. Detailed recommendations can be found in Wedemeyer (1966*a*).

Mineral salt formulations can also be used as water additives to mitigate stress and improve survival. The physiological benefits provided by either the simple or complex salt formulations that have been developed are mainly due to the protection they afford against the life-threatening blood electrolyte losses and ionoregulatory dysfunction that occur when the diuresis stimulated by handling and crowding stress is prolonged (Carmichael *et al.*, 1984). However, salt formulations can also mitigate other adverse physiological changes such as the depressed blood pH that results from metabolic acidosis in transported rainbow trout (Haswell *et al.*, 1982), and the hyperglycemia and hypercortisolemia that occur in transported anadromous salmonids (Wedemeyer, 1996*b*).

As judged from present experience, single or mixed formulations of NaCl, Ca(Cl)$_2$, Na$_2$SO$_4$, NaHCO$_3$, KCl, MgSO$_4$, K$_3$PO$_4$, and sea salts, used with or without tranquilizing concentrations of MS-222, have the most potential for alleviating the life-threatening physiological disturbances resulting from hauling smolts or other juvenile salmonids – especially in the slightly acidic waters of low total hardness typical of the west coast of North America (Wedemeyer, 1996b). It has been known for many years that survival rates of both cold-water and warm-water fish can be substantially increased by adding only NaCl at 0.5–1.0% to the hauling tank water. If fish can be allowed to recover in NaCl-enriched water after release from the hauling truck, survival is even better (Mazik, Simco & Parker, 1991). More complex mineral salt formulations have also been developed that are particularly useful in mitigating stress and reducing the mortality of fish transported in water of low total hardness (Carmichael et al., 1983, 1984).

Two disadvantages of mineral salt additives are (1) the potential for uptake of Mg^{2+} and K$^+$ if scale loss of more than 10% has occurred, and (2) the potential for increased equipment corrosion problems. Detailed recommendations for mineral salt formulations to mitigate stress and improve the survival of transported fish can be found in Wedemeyer (1996b).

Water additives containing polyvinylpyrrolidone (PVP) or other proprietary polymers can also be useful in fish transport operations. Abrasions and scale loss are common problems in transported juvenile and adult anadromous salmonids and have long been known to be a significant cause of delayed mortality after release. The cause of death is frequently a secondary fungal infection – a common occurrence when even minor skin abrasions disrupt the normal protective slime layer and expose the underlying tissue to attack by *Saprolegnia* spp. Polymer additives temporarily bond to the exposed tissue to form a protective coating which is gradually sloughed off as healing takes place and the normal slime layer re-forms. This technology has received only limited use in cold-water and warm-water fish culture where the approach has been to mitigate stress and improve survival by using transport additives such as mineral salts and anesthetics that mitigate physiological stress. However, polymer formulations, usually added as one of the commercially available products such as Polyaqua, are now being used increasingly by both the aquaculture industry and by state and federal conservation hatcheries as a water additive for transporting juvenile salmonids as well as other fish (Carmichael & Tomasso, 1988). In the author's experience, a concentration of 100 ppm Polyaqua during trucking has significantly reduced the prespawning mortality of adult fall (ocean-type) chinook salmon and steelhead trout due to *Saprolegnia.*

Withholding food for a short period of time prior to transportation is a technique used for improving the survival of transported fish that dates back to the 1800s (McCraren, 1978). Fasting decreases both the amount of oxygen consumed and the amount of carbon dioxide, ammonia, and feces produced. However, fish are evolutionarily adapted to intermittent feeding and it is now known that the metabolic rate may not decline significantly for several days. For example, the oxygen consumption of rainbow trout decreases by only about 25% and ammonia excretion by 50% after about 60 h of fasting. In general, salmonids smaller than about 0.1 g must be starved for at least 2 days before they are transported and larger fish for at least 3 days in order to lower their metabolic rates by an amount sufficient to produce a practical effect (Wedemeyer, 1996b).

The physiological benefits of transporting fish in chilled water have also been recognized by fish culturists since at least the 1800s (McCraren, 1978). Hypothermia reduces both oxygen consumption and the production of toxic excretory products. For warm-water fish, the Q_{10} (temperature coefficient) for physiological processes is approximately 2.0, and reducing the water temperature by 10 °C reduces the oxygen consumption and waste production of transported fish by about 50%. In salmonids, however, the Q_{10} for energy metabolism averages somewhat lower (1.4–1.9) than the expected value of 2.0 (Smith, 1977). Also, the relationship between energy metabolism and temperature is linear only over the range of 3–18 °C. Thus cooling the hauling tank water by only 8 °C decreases the oxygen consumption of transported rainbow trout by about 50% while lowering the water temperature by 6 °C reduces the oxygen consumption of transported sockeye salmon fry by only 20% (Wedemeyer, 1996b).

Hypothermia also assists in mitigating aspects of the physiological stress response. Lowering the water temperature from 10 °C to 5 °C reduces hyperglycemia by about 30% in spring chinook salmon smolts transported by truck at a loading density of 12% (0.12 kg/l, 1 lb/gal) (Wedemeyer, 1996b). Unfortunately, chilled water offers little protection against the blood electrolyte depletion that is a major factor in the delayed mortality which can occur following hauling.

Together with mitigating stress, chilled hauling tank water will allow the fish loading density to be increased (Piper et al., 1982). Thus, hauling tank water is usually cooled by 5–10 °C prior to loading both cold-water or warm-water fish as a matter of routine. Refrigeration systems or the use of chlorine-free ice to lower the water temperature to 5–7 °C are currently the most widely used methods in salmonid transport operations.

A mismatch of more than 10 °C between the hauling tank water and the receiving water can also be stressful to transported fish. Temperature shock can be minimized by acclimating the fish to the receiving water temperature over a period of several hours. The stress of temperature changes of less than 10 °C is mild and usually well tolerated by fish in good health. However, if the fish have been previously exposed to pathogens the mild stress response that does occur may still be sufficient to activate latent infections such as furunculosis or bacterial kidney disease (Wedemeyer & Goodyear 1984).

Reducing the light intensity is another technique that can be helpful in mitigating transport stress. Spring chinook salmon smolts subjected to a hauling challenge under darkened conditions show a 25% reduction in the hyperglycemia and hypochloremia normally caused by handling and crowding (Wedemeyer, 1996b). Except for behavioral avoidance, little is known about the physiological response of warm-water fish to brightness during transport operations. However, taking steps to reduce light intensity during all stages of smolt transport is a recommended practice.

Interactions between fish

The behavioral interactions that occur between fish in intensive aquaculture systems can be a significant source of stress. A number of behaviors have evolved in the interactions between fish in wild populations that enhance individual success in the endless competition for food and other resources such as space, favorable water quality, cover for protection against predators, and breeding partners. Under natural conditions, these behaviors are adaptive and serve to regulate life in groups. The behavior of most interest is aggression. Aggression is used to establish individual fish as territory holders, as dominants in intraspecific hierarchical social systems, and to enhance success in interspecific competition. Some aggressive behaviors, such as biting, charging, chasing, and ramming, can inflict serious physical injuries. Others, such as lateral displays, intimidate rivals without requiring direct physical contact. During the initial establishment of territories or social hierarchies, aggressive interactions tend to be frequent and intense. After social systems have been established, however, conflicts usually become less frequent and relatively stable relationships may develop between dominants and subordinates. Residency within established territories often becomes long term. As an alternative to permanent submission, less successful competitors usually also have the option of moving to another location. For example, fish can often find another feeding territory in an area where competition is less and frequent conflicts with dominants can be avoided. Of course, food availability may be lower and risk of predation higher.

When fish are reared under intensive conditions, such as in raceways with broadcast feeders, behaviors such as aggression do little to provide food or cover and simply become aberrant, causing unnecessary stress and physical injury. Migration as an alternative to frequent aggressive conflicts with dominants is also not an option. Broken or damaged fins, from repeated biting, and reduced growth due to chronic stress in the defeated individuals are common. Other behavioral interactions that adversely affect health and physiological condition include interspecific competition, social dominance hierarchies, and territoriality. Interspecific competition is usually not a consideration in intensive culture systems but can be a source of pathology in extensive polyculture or in experimentally mixed populations. For example, brook trout held with equal numbers of brown trout in artificial streams suffer a 33% mortality from *Saprolegnia* infections. Brown trout were never infected and neither were brook trout when they were held alone at the same density (Wald & Wilzbach, 1992).

Health problems resulting from social dominance in fish–fish interactions are a relatively common occurrence in aquaculture systems. Dominance hierarchies are based largely on fish size and can form relatively quickly in many species, particularly at low rearing densities. Fin nipping, scale loss from ramming, reduced growth, pathological changes in gastrointestinal tissue, and increased susceptibility to infectious diseases due to chronic physiological stress can all occur in defeated individuals. In eel culture, for example, subordinate fish in the dominance hierarchies that form show chronically elevated levels of corticosteroid (stress) hormones, atrophy of the gastric mucosa, and a suppressed immune protection system (Ejike & Schreck, 1980; Peters & Quang Hong, 1984). In Tilapia, leukocyte function is impaired in defeated fish (Cooper *et al.*, 1989).

The fact that dominated fish also grow more slowly reinforces size-based hierarchies and badly skewed or bimodal size distributions can develop in the ponds in a matter of weeks. The slower growth of subordinated fish in social hierarchies is due to a combination of reduced access to feed, the caloric energy costs of chronic stress, and the anorexia that occurs in chronically stressed animals in general. For example, subordinates in juvenile steelhead trout populations grow significantly more slowly than behavioral dominants even though food intake is experimentally equalized. Subordinates are also less active; presumably also in response to the caloric energy costs of the chronic stress (Abbot & Dill, 1989). The rationale for periodic size grading in hatcheries is that larger fish tend to be dominants and if they are removed growth of the remaining smaller fish usually improves. In addition, size grading does facilitate feeding because a smaller range of pellet sizes can be used to feed the graded groups. In some hatcheries, periodic size grading is routine because it is assumed to disrupt social hierarchies and thus reduce the divergent growth between dominant and subordinate individuals. In salmon hatcheries producing fish for conservation, size grading is sometimes used as a release strategy.

Size grading is not always beneficial. In less domesticated species, a greater degree of size variation may occur naturally because of their greater degree of genetic variability. In addition, although stream-dwelling species such as salmonids do show territorial and agonistic behavior under low density (natural) conditions, these behaviors are less prevalent under the high rearing densities used in intensive culture. In the schooling behavior that develops at high raceway densities, imitation rather than aggression is characteristic. This, together with the frequent or automatic feeding typical of intensive culture, tends to reduce size variation independently of grading (Wallace & Kolbeinshavn, 1988).

In addition to the use of aggression to form social systems, hatchery fish will also become aggressive if they are allowed the opportunity to compete for food. For example, juvenile Pacific salmon normally behave as loosely schooling fish under raceway rearing conditions. However, if they are fed from a point source (e.g., a single demand feeder), some individuals will begin to show aggressive rather than imitative behavior, and begin intimidating competitors and monopolizing food. As in dominance hierarchies, defeated individuals suffer ragged or broken fins from repeated biting and slower growth from stress-induced anorexia and reduced access to food.

The challenge for fish culture is to identify and provide rearing conditions under which behaviors such as aggression or social dominance are no longer effective in acquiring space, favorable water quality, food, or other resources. Theoretically, fish will only use aggressive behavior to obtain or defend food if its density and distribution, and the density and distribution of competitors, make aggression bioenergetically more profitable than using alternative tactics such as scrambling (Grant, 1993). In line with this, practical experience shows that skewed size distributions can be minimized by increasing the number of demand feeding stations per raceway, increasing the amount of feed released per trigger strike, or by using broadcast feeders. Under these conditions, it becomes easier for subordinates to take food not eaten by dominants or to simply wait until the dominants have fed to satiation and have dispersed. Thus, the energy cost of defeating other fish in order to monopolize food becomes too great and aggressive behavior tends to decline.

Aggressive behavior can also be modified by making adjustments to rearing densities. For tilapia, an important food fish that is aggressive, practical experience has shown that both aggressive and sexual behavior are density dependent and can be shifted toward the more desirable schooling behavior by simply increasing fish production levels. The agonistic behavior of most salmonids does tend to decrease at higher fish densities as does the establishment of dominant–subordinate hierarchies. This is not the case with all species, however, as shown by the fact that smaller eels may fail to feed if larger eels are present, even after the larger eels are satiated and food is still available (Knights, 1987). Even though the number of agonistic interactions initiated per individual may decline as the fish density is increased, total aggression in the raceway/pond may remain high as evidenced by the persistence of damaged fins, particularly in species such as steelhead. Nonetheless, behavioral considerations probably offer significant opportunities for health improvements in the production of most species by intensive culture methods.

It is also important to consider evolved behavioral tendencies when species are initially considered for intensive culture rearing. This will help to avoid the imposition of rearing conditions that inadvertently stimulate aggressive behavior or that exacerbate the effects of normal social conflicts. For example, lake-dwelling (schooling) species may be reared at densities far greater than those found in nature, with few problems. However, rearing a stream-dwelling territorial species at densities higher than those found in nature may provoke increased fin nipping and other aggressive acts by forcing continuous encounters between individuals. It is true that most species can adapt aspects of their behavior to a range of environmental conditions. However, forcing a strongly territorial fish to adopt nonterritorial, schooling, behavior, by raising the fish density so high that it precludes individual monopolization of available food, may simply cause other pathophysiological problems. Hatchery designs which incorporate no visual barriers to hinder direct eye contact between dominants and subordinates may also inadvertently stimulate aggressive behavior in some species. Feeding practices which stimulate aggression by delivering small amounts of food to small predictable areas have already been discussed.

The behavior modification that can occur in fish reared under intensive conditions may be undesirable if the fish are intended to supplement wild populations after release. For example, hatchery reared cutthroat trout that are transferred to artificial streams seem to be more aggressive than their wild conspecific counterparts, as evidenced by their greater use of fast flowing water and a higher incidence of lateral displays and biting. Hatchery-reared coho salmon also seem to be more aggressive than wild juveniles (Swain & Riddle, 1990). These behaviors appear to confer little survival advantage and the energy expenditure for the unnecessary aggression may be one factor that reduces success in the use of hatchery fish to supplement wild runs (Mesa, 1991). In contrast, when Atlantic salmon fry from hatchery stocks are released into artificial streams to compete for territories with wild fry, they are generally less aggressive. This may be because there is less selection pressure for dominance behavior in the rearing environments used in the intensive culture of these fish (Norman, 1987). Acclimation to the frequent episodes of stress (size grading, pond cleaning) encountered during hatchery rearing also seems to attenuate the natural physiological response to stressful challenges. If hatchery coho and chinook salmon are stocked as fry and allowed to acclimate to a natural stream environment for several months, they then show the same heightened cortisol response to stressful challenges as wild smolts (Salonius & Iwama, 1993).

The role of stress in infectious disease problems

Fish–microorganism interactions

The interactions between fish and the microorganisms present in the rearing environment are most often harmless or beneficial. However, these interactions take place in a dynamic environment to which both the fish and the microorganism are connected by complex ties (Kabata 1984). Thus, fish–microorganism interactions, as influenced by water quality factors and the stress of rearing conditions, can also result in infections and epizootic diseases.

Aquaculture personnel should have an understanding of the characteristics of fish pathogens because these characteristics can be used in disease control. In general, fish pathogens can be classed as being either obligate or facultative. Obligate pathogens cannot obtain the nutrients required to survive and remain infectious when suspended in the water column or attached to sediment particles and require a living host in order to grow and reproduce. However, when shed into the water by an infected host they can often remain viable long enough to be transmitted horizontally (fish to fish). This group of microorganisms includes all the viruses, some protozoan parasites (*Ichthyophthirius multifiliis*, *Chilodonella*, and *Ichthyobodo* spp., *Myxobolus cerebralis*), and a few bacteria such as *Aeromonas salmonicida* (furunculosis), *Yersinia ruckeri* (enteric red-mouth disease), and *Renibacterium salmoninarum* (bacterial kidney disease). Obligate pathogens will not normally be found free in a hatchery water supply unless aquatic animal life is also present to serve as a reservoir of infection. Thus, diseases caused by this group of microorganisms will not normally occur in hatchery fish unless other fish, amphibians or eggs infected with obligate pathogens also reside in, or are introduced into, the water supply. Personnel managing hatcheries that have springs or wells (closed water sources) can minimize or even prevent disease problems caused by obligate pathogens by simply removing any resident fish, frogs, or other amphibians that may be present in the water supply – provided that the fish are not already infected with pathogens that can be transmitted vertically. In hatcheries using water from open sources, such as small streams, it is sometimes possible to reduce obligate pathogen numbers by using fish screens to keep resident fish that are reservoirs of infection from moving above the hatchery intakes during their spawning migration. In either case, it is important to note that although obligate pathogens cannot live indefinitely after being shed into the water, they may still survive and remain capable of infecting fish for a number of days or even a few weeks depending on the water chemistry. Example survival curves for the obligate pathogens *A. salmonicida* (furunculosis), and the IHN (infectious hematopoietic necrosis) virus suspended in distilled, soft, or hard lake waters can be found in Wedemeyer (1996a).

Pathogens that do not require aquatic animal life forms as hosts but that can obtain sufficient nutrients to live and reproduce while attached to aquatic plants, suspended organic material (detritus), or even sediment particles are classed as facultative. Facultative pathogens are normally ubiquitous in natural waters.

In aquaculture, fish disease problems caused by facultative pathogens can be particularly troublesome. Examples in freshwater culture include bacteria such as *Aeromonas hydrophila* (motile *Aeromonas* septicemia), *Flexibacter columnaris* (columnaris disease), *Edwardsiella ictaluri* (enteric septicemia of catfish), many external protozoan parasites (*Trichodina* spp.), and fungi such as *Saprolegnia* spp. In marine net-pen culture *Vibrio anguillarum* (vibriosis) is a chronic problem. Since facultative pathogens are normally ubiquitous in surface waters, removing fish or other aquatic life forms will have little effect. Closed water supplies (springs, wells) that are free of aquatic animals, and thus obligate pathogens, are also generally free of facultative pathogens as well because of the extensive filtering of nutrients (carbon and nitrogen sources) by subterranean sands and gravel. Unfortunately, facultative pathogens will usually quickly recolonize the water as soon as it is exposed to the atmosphere in headboxes, tanks, ponds, or raceways.

Both classes of fish pathogens can be controlled by water treatment systems using ultraviolet (UV) light, ozone, chlorine, or other disinfectants. In hatcheries using recirculated water, or in smaller facilities using flow-through conditions, water treatment systems employing UV light or ozone may be practical. However, in pond culture, and in most large hatcheries operated under flow-through conditions, water treatment systems are usually impractical due to the high volumes or flow rates involved. Fortunately, many of the disease problems resulting from fish–microorganism interactions can also be minimized by managing rearing conditions to meet the physiological needs of the fish and to provide a low-stress environment. An example list of obligate and facultative fish pathogens together with the disease problems they cause is given in Table 2.

As mentioned, disease is not an inevitable outcome of the continuously occurring interactions between fish and aquatic microorganisms. However, the interactions between fish and microbial pathogens that may be harmless under natural conditions often result in disease problems in hatchery fish because of the added stress from the physical, chemical, or biological challenges inherent in intensive culture systems. Such challenges elicit a catecholamine and corticosteroid hormone cascade that brings about a series of cardiovascular, respiratory, and other secondary physiological changes intended to help the fish avoid or escape from the challenge in question. However, in aquaculture systems opportunities for avoidance or escape are minimal, the physiological changes usually do little to improve survival, and may become harmful if prolonged. The harmful effects include varying degrees of growth suppression, reproductive dysfunction, and immunosuppression. The immunosuppression is particularly important because its effects can linger for some time after the other physiological changes have returned to prestress levels (Maule *et al.*, 1989).

Many of the physiological mechanisms that allow infections to become established and to spread when fish are stressed are not understood, but the immunosuppression is apparently a side-effect of cortisol pro-

Table 2. *Examples of facultative and obligate pathogens important to intensive fish culture*[a]

Facultative pathogens	Obligate pathogens
Aeromonas hydrophila (motile *Aeromonas* septicemia)	Aeromonas salmonicida (furunculosis)
Vibrio anguillarum (vibriosis)	All fish viruses (IHN, CCV, VHS, IPN)
Flexibacter columnaris (columnaris disease)	*Flavobacterium branchiophilum* (bacterial gill disease)
Edwardsiella ictaluri (enteric septicemia of catfish)	Protozoan parasites (*Ichthyobodo*, *Chilodonella*, *Ichthyophthirius*, *Myxobolus cerebralis*)
Fungi (*Branchiomyces* spp., *Saprolegnia* spp.)	*Renibacterium salmoninarum* (bacterial kidney disease)
Vibrio salmonicida (Hitra disease of farmed Atlantic salmon)	*Yersinia ruckeri* (enteric red-mouth disease)

[a]After Wedemeyer (1996*a*).
Notes: *Flavobacterium branchiophilum* originally thought to be a facultative pathogen, *Edwardsiella ictaluria* originally thought to be obligate.
IHN, infectious hematopoietic necrosis.

duction. The immune response to invading pathogens can also be adversely affected by nutritional deficiencies, but stress factors such as biological, chemical, and physical challenges are probably more important to fish in intensive culture systems. The specific and nonspecific aspects of the immune protection system can both be compromised.

Serum bactericidal activity, due to lysozyme and the complement system, and phagocytosis by macrophages and other leukocytes are the main features of the nonspecific immune response in fish. All of these features are adversely affected when fish are stressed. For example, carp subjected to a low DO concentration, starvation, or high-salinity water suffer a significant decrease in serum bactericidal activity. The complement system and lysozyme activity are both depressed (Hajji *et al.*, 1990). Lysozyme activity in rainbow trout remains low for at least 24 h after the fish have been transported for only 2 h (Moeck & Peters, 1990).

Macrophage phagocytosis is impaired by a variety of stress factors. In channel catfish and rainbow trout, these include handling, transportation, confinement and even noise (Ellsaesser & Clem, 1986; Nanaware, Baker & Tomlinson, 1994). In most cases, both the ability to phagocytize (initial engulfment) and the intracellular killing of the phagocytized bacteria are adversely affected. The weakened ability to phagocytize initially is apparently due to the direct action of increased blood cortisol concentrations on the fluidity of the macrophage cell membrane. The failure to kill bacteria after they have been phagocytized is apparently also a side-effect of hypercortisolemia. Two mechanisms are probably involved: failure of intracellular enzymes to disintegrate the ingested microbe after it is fused with the cytoplasmic phagosome, and interference with the killing action of nonspecific cytotoxins produced by the macrophage. The cytotoxins include hydrogen peroxide (H_2O_2) and the superoxide anion (O_2^-). Both are produced from oxygen during the respiratory burst that is stimulated when microbes attach to the plasma membrane during the phagocytic process. The macrophage protects itself from damage by destroying excess amounts of these highly reactive oxygen species with the enzymes catalase and superoxide dismutase located in the cytoplasm. Bacteria that are phagocytized but not killed can continue to metabolize and divide and will eventually kill the macrophage. In fish, effects on disease resistance seem to involve failure of the intracellular killing process more than failure of phagocytosis itself. For example, macrophages containing viable and apparently dividing bacteria can often be found in salmonids suffering from bacterial kidney disease (*Renibacterium salmoninarum*) or channel catfish with *Edwardsiella* septicemia (*E. ictaluri*) infections (Blazer, 1991).

Effects of stress on other important macrophage functions, such as debris removal during tissue repair, uptake and processing of antigens, and the secretion of lysozyme and interferon, may also be involved in allowing infections to become established and spread but are little understood at present. Evidence for the role of interferon in fish is indirect, but in rainbow trout, for example, recovery from VHS (viral hemorrhagic septicemia) infections often occurs spontaneously if the water temperature is increased to 15 °C or more. However, the virus develops quite normally *in vitro* at these temperatures. The thermodependence of the clinical disease may be partly due to the protective effects of faster interferon synthesis by leukocytes at warmer temperatures (de Kinkelin, Dorson & Renault, 1992).

Together with impaired phagocyte function, total numbers of circulating leukocytes are also reduced in stressed fish. The leukopenia (mainly lymphocytopenia) is also a side-effect of hypercortisolemia and plays a major role in increasing susceptibility to microbial invasion and in allowing infections to spread (Pickering 1984; Barton & Iwama, 1991).

In addition to decreasing the circulating leukocyte count, stress-induced hypercortisolemia also interferes with antibody production by lymphocytes. For example, Atlantic salmon parr subjected to the stress of confinement for only 2 h suffer impaired lymphocyte function and reduced production of specific antibody to *A. salmonicida* (furunculosis) (Thompson *et al.*, 1993). The mechanism apparently involves suppression of cytokines such as interleukin-1 that are involved in antibody production (Kaattari & Trip, 1987). However, it has long been recognized that serum antibody titer does not necessarily correlate well with disease resistance in fish – probably because of their greater reliance on the nonspecific mechanisms of the immune system for protection against invading microorganisms (Landolt, 1989).

As mentioned, the mechanisms involved in the establishment and spread of infections in stressed fish also include a link to nutritional status. As might be expected, the naturally occurring interactions between nutritional deficiencies, the immune system, and infectious agents are complex and studies have often yielded conflicting results. Nonetheless, a few general principles have been revealed.

Frank malnutrition usually facilitates infections because of general debilitation of the physiological condition and the immune system. For example, starvation increases the prevalence of ceratomyxosis (*Ceratomyxa labracis* and *C. diplodae*) in cultured sea bass (Alvarez-Pellitero & Sitjà-Bobadilla, 1993). One mechanism may be that starvation tends to reduce the nonspecific bactericidal activity of serum complement and lysozyme. Malnutrition can result in antagonism of infection if the infectious agent is dependent on specific host enzyme systems or metabolites, or has a greater requirement than the host for a particular dietary nutrient.

Vitamin and micronutrient deficiencies usually act in synergy in infections. Vitamin C, in particular, is widely considered to have generally beneficial effects on disease and stress resistance in both salmonids and channel catfish when fed at the basic requirement level of 50–100 ppm (mg/kg). Feeding megadoses of vitamin C (3000 mg/kg) completely protected channel catfish against *Edwardsiella ictaluri* (enteric septicemia) infections (Li & Lovell, 1985). Feeding vitamin C (and E) at higher than basic requirement levels has also been shown to have additional beneficial effects on the disease resistance of Atlantic and Pacific salmon (Leith & Kaattari, 1989; Hardie, Fletcher & Secombes, 1990). Unfortunately, specific biochemical and physiological mechanisms by which vitamins and micronutrients act to affect disease resistance have been difficult to elucidate. For example, leukocyte bactericidal activity, migration, and phagocytic index (mean numbers of bacteria phagocytized per cell), are all unaffected in salmonids and channel catfish fed high levels of vitamin C (Johnson & Ainsworth, 1991; Thompson et al., 1993).

The vitamin C status and nutritional state can also influence the spread of infections by affecting production and maintenance of repair tissue. Vitamin C and the sulfur-containing amino acids are required for the deposition of fibrin, collagen, and polysaccharides into the reticulum formed to isolate invading microbial pathogens by walling-off infections. Deficiencies can inhibit the walling-off process. In addition, the walling-off process, especially the polysaccharide and protein (fibrin) components of the reticulum, is also susceptible to hydrolytic and proteolytic enzyme attack. A substantial increase in nonspecific plasma proteolytic activity can be stimulated by bacterial pathogens that produce endotoxins, or by certain types of stressful challenges. Chronically stressful rearing conditions such as a low DO concentration, tend to decrease serum lysosome activity, while acute challenges such as transportation or confinement increase it in both carp and Atlantic salmon (Hajji et al., 1990; Thompson et al., 1993). Thus, it is possible that acute stress may act synergistically with a vitamin C deficiency to facilitate the spread of invading pathogens through fish tissues.

Vitamin B complex and mineral deficiencies behave variably depending on the host and infectious agent. Pyridoxine (vitamin B_6 group) improves the resistance of salmonids to Vibrio anguillarum (vibriosis), Aeromonas salmonicida (furunculosis), and Renibacterium salmoninarum (bacterial kidney disease or BKD) infections (Hardy, Halver & Brannon, 1979; Leith & Kaattari, 1989). In the case of mineral deficiencies, low serum iron levels would be expected to reduce the invasiveness of bacterial pathogens such as V. anguillarum (vibriosis), R. salmoninarum (BKD), and A. salmonicida (furunculosis) that use iron-chelating siderophores to obtain the serum iron needed for their pathogenicity. Thus, farmed Atlantic salmon with high levels of serum iron are more susceptible to Vibrio infections compared to fish with lower iron levels (Ravndal et al., 1994). In contrast, dietary iron, zinc, or iodine deficiencies seem to increase the susceptibility of farmed Atlantic salmon to BKD (Plumb, 1994). Micronutrients may also affect resistance to infections by affecting the process of phagocytosis. This could play a critical role in fish because of their greater reliance on the nonspecific defenses of the immune protection system. For example, copper and zinc are required for optimum activity of superoxide dismutase, the enzyme that helps protect phagocytic cells against damage by the cytotoxins they use to kill engulfed microorganisms. No work has been done with fish, but zinc deficiency in rats results in superoxide, free radical, and H_2O_2 accumulation in several cell phagocytic cell types (Kubow, Bray & Bettger, 1986).

Dietary lipid composition may also affect macrophage function. Channel catfish fed diets formulated with fish oils rich in omega-3 unsaturated fatty acids produced macrophages that had much higher bactericidal activity than those from fish fed diets formulated with lipids high in saturated fat (Blazer, 1991). One mechanism for this may be that the fatty acid composition of the macrophage cell membrane, which influences its fluidity, is easily altered by diet (Johnston, 1988). The possibility of enhancing macrophage function by manipulating diet constituents is an attractive concept for managing the health of fish, although it is a remote possibility at present.

Stress-mediated diseases

Snieszko (1954) was among the first to point out that the incidence of certain freshwater fish diseases in Europe and North America tended to be seasonal, while no such relationship usually existed in tropical climates. That is, environmental stress (seasonal water temperature changes), in addition to exposure to pathogen, was required for the initiation of epizootics. In succeeding years, it became clear that problems with bacterial diseases tended to increase in the spring because the growth and invasiveness of the pathogens involved responded more rapidly to the rising temperatures of the water than the immune defenses of the fish whose physiological condition had been debilitated by the overwintering environment (Wedemeyer, Meyer & Smith, 1976). Since then, it has become well recognized that most of the infectious disease problems of hatchery (or wild) fish are multifactorial in nature and not simply the inevitable result of exposure to pathogens as previously supposed.

Because of the complexity of the interactions between the pathogen, the aquatic environment, and the physiological condition of the fish, not all the specifics of the physical, chemical, and biological stress factors associated with outbreaks have been defined. For fish in intensive culture, however, the commonly occurring predisposing factors have been identified. These include stressful hatchery practices such as handling, size grading, crowding, and transportation under adverse water quality conditions, for example a low DO concentration, high ammonia and carbon dioxide concentrations, and unfavorable temperatures. The chief pathogens involved in the stress-mediated diseases of hatchery fish include the external fungi, protozoan parasites, most of the facultative and obligate bacteria, and a few viruses. Detailed information about the role of environmental factors in the epizootiology of most microbial diseases can be found in Plumb (1994). A few of particular interest are discussed here.

Of all the stress-mediated fish disease problems identified to date, those due to facultative bacterial pathogens are probably the most frequently encountered. Diseases due to obligate pathogens can be stress mediated as well, but are somewhat less frequent because these microorganisms are not always present in the rearing environment. The usual stress factors involved are fish culture practices and unfavorable water quality conditions. Infections such as the motile *Aeromonas* septicemias (*A. hydrophila* and others), bacterial gill disease (*Flavobacterium branchiophilum*), furunculosis (*A. salmonicida*), columnaris (*Flexibacter columnaris*), and vibriosis (*Vibrio anguillarum*) are classic examples of stress-mediated bacterial diseases (Wedemeyer & Goodyear, 1984).

Epizootics of motile *Aeromonas* septicemia (MAS), a hemorrhagic septicemia caused by *A. hydrophila* and other motile species of the genus *Aeromonas*, have long been a ubiquitous problem in both cold-water and warm-water fish culture. In warm-water pondfish culture, *A. hydrophila* is considered to be one of the primary bacterial pathogens; its major occurrence on catfish farms is during June, July, and August corresponding to the season of the year when pond conditions are least favorable for fish. DO depletions are common during this period and water temperatures are rising. MAS infections in channel catfish usually do not occur unless their physiological condition is debilitated by multiple stress factors (Plumb, 1994). The fish culture procedures identified as contributory include overcrowding, overfeeding, excessive pond fertilization, and algal die-offs after blooms (Walters & Plumb, 1980). The usual water quality factors involved are a low DO level (1–2 mg/l), elevated carbon dioxide (5–10 mg/l), and elevated ammonia (1–2 mg/l). Disease problems are often delayed for 1–2 weeks after the fish are stressed by a combination of adverse water quality conditions and stressful fish culture procedures (Plumb, 1994).

In cold-water raceway culture, the most common stress factors are handling and crowding. However, in rainbow trout, an important cold-water fish which is aggressive, the stress of behavioral conflicts alone can be sufficient to result in MAS in defeated individuals if the pathogen is present in the water (Peters *et al.*, 1988).

Furunculosis, a bacterial septicemia of salmonids and a few cool-water and warm-water fish, is another disease often correlated with stressful water quality conditions and fish culture procedures. The etiological agent *A. salmonicida* is an obligate pathogen but usually disease outbreaks do not occur independently of stressful conditions. Kingsbury (1961) was the first to correlate systematically furunculosis outbreaks in hatchery rainbow trout with specific environmental conditions – water temperatures above 10 °C, DO levels below 5.5–6.0 mg/l, handling for size grading and transportation, and crowding beyond species guidelines. In Pacific salmon, crowding in terms of the areal density (number of fish/ m^2) as well as the related volume density (number of fish/m^3) can also be important (Nomura, Yoshimizu & Kimura, 1992). For example, the incidence of *A. salmonicida* infections in adult chum salmon held in shallow freshwater ponds at areal densities of about 15 fish/m^2 was about 12%, while fish held at the reduced density of 5 fish/m^2 were disease free. Similarly, the prevalence of *A. salmonicida* in organs and tissues was significantly higher in adult chum salmon held at the marginal DO levels of 6–7 mg/l than in fish held in water with a DO level of 10 mg/l.

Managing rearing conditions to maintain the water temperature below 15 °C, the DO level near saturation, and fish loadings within species guidelines, as well as minimizing handling during high-risk periods when water temperatures are rising are all helpful in preventing furunculosis outbreaks. Water temperatures of 15–20 °C exacerbate furunculosis problems in salmonids. If the underlying environmental conditions are corrected, drug treatments will suppress outbreaks. However, surviving fish normally become *A. salmonicida* carriers. Smolts carrying this pathogen will often have the latent infection activated by the stress of transfer into seawater. The resulting reduced early marine survival can cause serious economic losses. In the commercial aquaculture of Pacific and Atlantic salmon, the corticosteroid–heat stress test is sometimes used to identify smolts free of latent infections prior to transfer into seawater netpens for grow-out (Eaton, 1988).

Bacterial gill disease (BGD) is a secondary infection by loosely defined filamentous bacteria that usually does not occur unless predisposing environmental conditions have first irritated the gill tissue. The principal infectious agent is *Flavobacterium branchiophilum* (syn. *branchiophila*), but *Cytophaga* and *Flexibacter* spp. are also known to infect injured gill tissue. Masses of these bacteria growing on the gill surface fatally impede gas exchange. BGD is primarily a disease of intensively cultured freshwater salmonids but warm-water fish including carp, catfish, and eels can also be affected. It has long been accepted that subtle gill injuries caused by elevated ammonia levels, suspended feed particles, or turbidity from suspended sediments are the main predisposing factors to bacterial invasion. In intensive culture systems, most outbreaks of BGD are associated with management errors such as overfeeding, overcrowding, inadequate water flow rates, low DO levels, increased un-ionized ammonia levels, and the accumulations of suspended particulate matter that causes gill irritation. However, the factors in the host–pathogen–environment relationships that predispose fish to BGD seem to be more complex than for other stress-mediated diseases. BGD occasionally occurs in fish held under rearing conditions that appear to be optimal. As a consequence, BGD has been difficult to reproduce under controlled laboratory conditions even when the experimental fish are severely stressed.

If the underlying environmental problems can be corrected, fish with BGD recover rapidly when the bacteria are removed from their gills with treatment chemicals. The standard therapeutants are compounds containing quaternary ammonium salts [e.g., Hyamine (benzethonium chloride) or Roccal (benzalkonium chloride)] or Chloramine-T (sodium p-toluenesulfonchloramide). A necessity for repeated treatments is evidence that the underlying environmental stress factors have not been correctly identified.

Other common bacterial fish diseases for which there is a high risk of outbreaks when unfavorable water temperatures and/or other stressful environmental conditions occur include enteric red-mouth (*Yersinia ruckeri*) in rainbow trout and enteric septicemia (*Edwardsiella ictaluri*) in catfish. For example, the risk of enteric red-mouth outbreaks is highest during the spring and summer when water temperatures are between 11 and 18 °C (Romalde *et al.*, 1994). Outbreaks of enteric septicemia of catfish (ESC), the leading bacterial disease of farmed channel catfish in the United States of America, are also related to environmental conditions; however, the relationship is not well understood. ESC generally occurs seasonally (late spring, fall) when water temperatures are in the 18–28 °C range, suggesting a cause and effect relationship, but outbreaks can also occur when environmental conditions appear to be highly favorable to the fish (Plumb, 1994). Handling, transportation, and confinement stress have all been shown to increase the mortality rate (Wise, Schwedler & Otis, 1993). The cost of vaccines, and the recent development of *E. ictaluri* strains resistant to the only drug presently labeled for ESC control (Romet, sulfadimethoxine potentiated with ormetoprim) are persuasive reasons for developing a better understanding of the ESC–aquatic environment relationship.

Seasonal changes in water temperature also strongly affect the incidence of diseases due to parasite infestations. In warm-water pond culture, oxygen depletions and high dissolved carbon dioxide concentrations due to the uncontrolled growth of aquatic macrophyte plants are particularly likely to encourage infestations due to parasites such as *Argulus*, *Gyrodactylus*, and *Trichodina* spp. Managers must therefore recognize the optimal environmental ranges for each parasite and take preventive action through prophylactic formalin treatments or be prepared to treat as soon as infections are identified. *Ichthyophthirius*, *Gyrodactylus*, and *Lernaea* provide particularly graphic examples of the relationship between infection and temperature.

The incidence of *Ichthyophthirius multifiliis* in warm-water pond-fish culture shows a distinct peak during the spring months, with a marked prevalence during April. Low water temperatures during the winter retard development of an epizootic until spring when warming trends begin. Water temperatures begin to climb in April and can approach the optimal range of 21–24 °C, which favors the development of this parasite. Handling stresses during April may also contribute to outbreaks in catfish. It is also possible that the April peak may reflect late recognition of the disease by fish culturists, since this parasite is probably present in low numbers all during the winter.

Gyrodactylus spp. epizootics on channel catfish farms in North America also show a peak occurrence during April although outbreaks are not unusual from January until July. Water temperature probably plays its most important role in determining the generation time of *Gyrodactylus* spp. At 20–24 °C, only a few days are required between the time that eggs are produced and hatched and the larvae locate a host fish and develop into adults. At water temperatures below about 10 °C this time may be extended to several months (Post, 1987).

The season for *Lernaea* parasitism is when water temperatures rise above 18 °C, the optimum being 22–30 °C. Warm weather in the early spring or late fall may extend the period of infection beyond the normal season of April to October.

A number of noninfectious (physiological) fish disease problems also occur as the result of improper management of rearing conditions. Common examples in freshwater include environmental gill disease (gill necrosis), coagulated yolk disease of salmonid eggs and fry, swimbladder stress syndrome, and the life-threatening diuresis responsible for delayed mortalities after fish are transported (Klontz, 1993; Wedemeyer & Goodyear, 1984). The so-called swim bladder stress syndrome can be a serious physiological problem in the intensive rearing of tilapia, striped bass, and marine perciform fish such as the sea bream. Physostomous and some physoclistic fish initially inflate the gas bladder by swallowing air at the surface. Larval fish may be prevented from doing this by the oil film left

Table 3. *Environmental factors commonly associated with the occurrence of infectious and noninfectious fish diseases*[a]

Fish disease problem	Predisposing environmental factors
Bacterial gill disease (*Flavobacterium* sp.)	Crowding; chronic low oxygen (4 mg/l for salmonids); elevated ammonia (more than 0.02 mg/l for salmonids); suspended particulate matter
Blue sac, hydrocele	Temperature; ammonia; crowding
Columnaris (*Flexibacter columnaris*)	Crowding or handling during warm-water periods if carrier fish are present
Environmental gill disease	Adverse rearing conditions, but contributory factors currently not well defined
Epithelial tumors, ulceration	Chronic, sublethal contaminant exposure
Fin erosion	Crowding; low level of dissolved oxygen; nutritional imbalances; chronic exposure to trace contaminants; high total suspended solids; secondary bacterial invasion
Furunculosis (*Aeromonas salmonicida*)	Low oxygen (<5 mg/l for salmonids); crowding; temperature; handling when pathogen carriers are present
Hemorrhagic septicemias, red-sore disease (*Aeromonas, Pseudomonas*)	External parasite infestations; ponds not cleaned; crowding; elevated ammonia; low oxygen; stress due to elevated water temperatures; handling after overwintering at low temperatures
Kidney disease (*Renibacterium salmoninarum*)	Water hardness less than about 100 mg/l (as $CaCO_3$); diet composition; crowding; temperature
Nephrolithiasis	Water high in phosphates and carbon dioxide
Parasite infestations	Overcrowded fry and fingerlings; low oxygen; excessive size variation among fish in ponds

Table 3. (*cont.*)

Fish disease problem	Predisposing environmental factors
Skeletal anomalies	Chronic, sublethal contaminant exposure; adverse environmental quality; PCB, heavy metals, kepone, toxaphene exposure; dietary vitamin C deficiency
Spring viremia of carp	Handling after overwintering at low temperatures
Strawberry disease (rainbow trout)	Uneaten feed; fecal matter with resultant increased saprophytic bacteria; allergic response
Sunburn	Inadequately shaded raceways; dietary vitamin imbalance may be contributory
Swim bladder stress syndrome	Oil films; hypoxia; salinity; other water quality factors
Vibriosis (*Vibrio anguillarum*)	Handling; oxygen <6 mg/l, especially at water temperatures of 10–15 °C; salinity 10–15‰
White-spot, coagulated-yolk disease	Environmental stress: supersaturation >102–103%, temperature, metabolic wastes, chronic trace contaminant exposure

[a]After Wedemeyer (1996*a*).
Note: Incidence of bacterial kidney disease may be inversely proportional to ionic composition, but not necessarily to total hardness (Fryer & Lannan, 1993).

on the surface by some formulated diets. Hypoxic conditions may prevent initial inflation in physoclistic fish, such as Tilapia, that do not initially swallow air at the surface (Summerfelt, 1991). In marine fish culture, swimbladder stress syndrome is an intermittent physiological problem in the intensive rearing of sea bass and sea bream. The survival rate of larval fish with underinflated swim bladders is very low if they are challenged by any kind of stress such as handling or low oxygen (Soares, Dinis & Pousao-Ferreira, 1994).

A summary of the common infectious and noninfectious fish disease problems currently thought to be associated with adverse rearing conditions is presented in Table 3.

References

Abbott, J.C. & Dill, L.M. (1989). The relative growth of dominant and subordinate juvenile steelhead trout (*Salmo gairdneri*) fed equal rations. *Behavior*, **108**, 104–11.

Alvarez-Pellitero, P. and Sitjà-Bobadilla, A. (1993). *Ceratomyxa* spp. (Protozoa: Myxosporea) infections in wild and cultured sea bass, *Dicentrarchus labrax*, from the Spanish Mediterranean area. *Journal of Fish Biology*, **42**, 889–901.

Banks, J.L., Taylor, W.G. & Leek, S.L. (1979). *Carrying Capacity Recommendations for Olympia Area National Fish Hatcheries.* Abernathy Hatchery Technology Development Center, US Fish and Wildlife Service, Washington DC.

Barton, B.B. & Iwama, G.K. (1991). Physiological changes in fish from stress in aquaculture with emphasis on the response and effects of corticosteroids. *Annual Review of Fish Diseases*, **1**, 3–26.

Basu, S.P. (1959). Active respiration of fish in relation to ambient concentrations of oxygen and carbon dioxide. *Journal of the Fisheries Research Board of Canada*, **16**, 175–212.

Blazer, V.S. (1991). Piscine macrophage function and nutritional influences: a review. *Journal of Aquatic Animal Health*, **3**, 77–86.

Carmichael, G.J. & Tomasso, J.R. (1988). Survey of fish transportation equipment and techniques. *Progressive Fish-Culturist*, **50**, 155–9.

Carmichael, G.J., Wedemeyer, G.A., McCraren, J.D. & Millard, J.L. (1983). Physiological effects of handling and hauling stress on smallmouth bass. *Progressive Fish-Culturist*, **45**, 110–13.

Carmichael, G.J., Tomasso, J.R., Simco, B.A. & Davis, K.B. (1984). Characterization and alleviation of stress associated with hauling largemouth bass. *Transactions of the American Fisheries Society*, **113**, 778–85.

Ceballos, J.R., Petit, S.W. & Mckern, J.L. (1991). Fish transportation oversight team annual report – FY 1990. *Transport Operations on the Snake and Columbia Rivers*. NOAA Technical Memorandum NMFS F/NWR-29. U.S. Department of Commerce, National Oceanic and Atmospheric Administration, National Marine Fisheries Service, Washington DC.

Colt, J. & Orwicz, K. (1991). Modeling production capacity in aquatic culture systems under freshwater conditions. *Aquacultural Engineering*, **10**, 1–29.

Collins, C.M., Burton, G.L. & Schweinforth, R.L. (1987). *Evaluation of Liquid Oxygen to Increase Channel Catfish Production in Heated Water Raceways*. Tennessee Valley Authority, TVA/ONRED/AWR-8418. Gallatin, Tennessee.

Cooper, E.L., Peters, G., Ahmed, I., Faisal, M. & Choneum, M. (1989). Aggression in Tilapia affects immunocompetent leucocytes. *Aggressive Behavior*, **15**, 13–22.

Eaton, C.A. (1988). The use of stress testing to prevent the movement of salmonids that are latently infected with furunculosis. *Bulletin of the Aquaculture Association of Canada*, **88**, 73–5.

Ejike, C. & Schreck, C.B. (1980). Stress and social hierarchy rank in coho salmon. *Transactions of the American Fisheries Society*, **109**, 423–6.

Ellsaesser, C.F. & Clem, L.W. (1986). Haematological and immunological changes in channel catfish stressed by handling and transport. *Journal of Fish Biology*, **28**, 511–21.

Fryer, J.L. & Lannan, C.N. (1993). The history and current status of *Renibacterium salmoninarum*, the causative agent of bacterial kidney disease in Pacific salmon. *Fisheries Research*, **17**, 15–33.

Grant, J.W.A. (1993). Whether or not to defend? The influence of resource distribution. *Marine Behavior and Physiology*, **23**, 137–53.

Hajji, N., Sugita, H., Ishii, S. & Deguchi, Y. (1990). Serum bactericidal activity of carp (*Cyprinus carpio*) under supposed stressful rearing conditions. *Bulletin of the College of Agriculture and Veterinary Medicine*, **47**, 50–4.

Hardie L.J., Fletcher, T.C. & Secombes, C.J. (1990). The effect of vitamin E in the immune response of the Atlantic salmon (*Salmo salar*). *Aquaculture*, **87**, 1–13.

Hardy, R.W., Halver, J.E. & Brannon, E.L. (1979). The effect of dietary pyridoxine levels on growth and disease resistance of chinook salmon. In *Finfish Nutrition and Fish Feed Technology*, Volume I. Halver, J. & Tiews, K. (eds.) pp. 253–60. Heenemann, Berlin.

Haswell, M.S., Thorpe, G.J., Harris, L.E., Mandis, T.C. & Rauch, R.E. (1982). Millimolar quantities of sodium salts used as prophylaxis during fish hauling. *Progressive Fish-Culturist*, **44**, 179–82.

Hobe, H., Wood, C.M. and Wheatly, M.G. (1984). The mechanisms of acid–base and ion regulation in the freshwater rainbow trout during environmental hyperoxia and subsequent normoxia. 1. Extra- and intracellular acid–base status. *Respiratory Physiology*, **55**, 139–54.

Johnson, M.R. & Ainsworth, A.J. (1991). An elevated level of ascorbic acid fails to influence the response of anterior kidney neutrophils to *Edwardsiella ictaluri* in channel catfish. *Journal of Aquatic Animal Health*, **3**, 266–73.

Johnston, P.V. (1988). Lipid modulation of immune responses. In *Nutrition and Immunology*. Chandra, R.K. (ed.) pp. 37–86. Alan R. Liss, New York.

Kaattari, S.L. & Tripp, R.A. (1987). Cellular mechanisms of glucocorticoid immunsuppression in salmon. *Journal of Fish Biology*, **31** (Supplement A), 129–32.

Kabata, Z. (1984). Diseases caused by metazoans: crustaceans. In *Diseases of Marine Animals*, Volume 4, Part 1. Kinne, O. (ed.)

pp. 321–99. Introduction, Pisces. Biologische Anstalt Helgoland, Hamburg.

Kebus, M. J, Collins, M.T., Brownfield, M.S., Amundson, C.H., Kayes, T.B. & Malison, J.A. (1992). Effects of rearing density on the stress response and growth of rainbow trout. *Journal of Aquatic Animal Health*, **4**, 1–6.

Kingsbury, O.R. (1961). A possible control of furunculosis. *Progressive Fish-Culturist*, **23**, 136–8.

de Kinkelin, P., Dorson, M. & Renault, R. (1992). Interferon and viral interference in viroses of salmonid fish. In *Proceedings of the Oji International Symposium on Salmonid Diseases*. Kimura, T. (ed.), pp. 241–9. Hokkaido University Press, Sapporo, Japan.

Klontz, G.W. (1993). Environmental requirements and environmental diseases of salmonids. In *Fish Medicine*. Stoskopf, M. (ed.) pp. 333–42. W.B. Saunders, Philadelphia, Pennsylvania.

Knights, B. (1987). Agonistic behavior and growth in the European eel, *Anguilla anguilla* L, in relation to warm-water aquaculture. *Journal of Fish Biology*, **31**, 265–76.

Kubow, S., Bray, T.M. & Bettger, W.J. (1986). Effects of dietary zinc and copper on free radical production in rat lung and liver. *Canadian Journal of Physiology and Pharmacology*, **64**, 1281.

Lall, S.P., Paterson, W.D., Hines, J.A. & Hines, N.J. (1985). Control of bacterial kidney disease in Atlantic salmon *Salmo salar* L, by dietary modification. *Journal of Fish Diseases*, **8**, 113–24.

Landolt, M.L. (1989). The relationship between diet and the immune response of fish. *Aquaculture*, **79**, 193–206.

Leith, D. & Kaattari, S. (1989). *Effects of Vitamin Nutrition on the Immune Response of Hatchery-Reared Salmonids*. Final Report US Department of Energy, Bonneville Power Administration, Division of Fish and Wildlife, Portland, Oregon.

Li, Y. & Lovell, R.T. (1985). Elevated levels of dietary ascorbic acid increase immune responses in channel catfish. *American Journal of Nutrition*, **115**, 123–31.

Maule, A.G., Tripp, R.A., Kaattari, S.L. & Schreck, C.D. (1989). Stress alters immune function and disease resistance in chinook salmon (*Oncorhynchus tshawytscha*). *Journal of Endocrinology*, **120**, 135–42.

Mazik, P.M., Simco, B.A. & Parker, N.C. (1991). Influence of water hardness and salts on survival and physiological characteristics of striped bass during and after transport. *Transactions of the American Fisheries Society*, **120**, 121–6.

Mazur, C.F. & Iwama, G.K. (1993). Handling and crowding stress reduces the number of plaque forming cells in Atlantic salmon. *Journal of Aquatic Animal Health*, **5**, 98–101.

McCraren, J.P. (1978). History. In *Manual of Fish Culture, Section G: Fish Transportation*. Smith, C. (ed.) pp. 1–6. US Fish and Wildlife Service, Washington, DC.

Meade, J.W. (1991). Application of the production capacity assessment bioassay. In *Fisheries Bioengineering Symposium, American Fisheries Society Symposium 12*. Colt, J. & White, R. (eds.) pp. 365–7. American Fisheries Society, Bethesda, Maryland.

Mesa, M.G. (1991). Variation in feeding, aggression, and position choice between hatchery and wild cutthroat trout in an artificial stream. *Transactions of the American Fisheries Society*, **120**, 723–7.

Moeck, A. & Peters, G. (1990). Lysozyme activity in rainbow trout, *Oncorhynchus mykiss* (Walbaum), stressed by handling, transport, and water pollution. *Journal of Fish Biology*, **37**, 873–85.

Nanaware, Y.K., Baker, B.I. & Tomlinson, M.G. (1994). The effect of various stresses, corticosteroids, and antigenic agents on phagocytosis in the rainbow trout *Oncorhynchus mykiss*. *Fish Physiology and Biochemistry*, **13**, 31–40.

Norman, L. (1987). *Stream Aquarium Observations of Territorial Behavior in Young Salmon (Salmo salar) of Wild and Hatchery Origin*. Report of the Salmon Research Institute (Laxforskningsinstitute), Sweden.

Nomura, T., Yoshimizu, M. & Kimura, T. (1992). An epidemiological study of furunculosis in salmon propagation. In *Salmonid Diseases*. Kimura, T. (ed.) pp. 187–93. Hokkaido University Press, Hokodate, Japan.

Pennell, W. 1991. *Fish Transportation Handbook*. Province of British Columbia, Ministry of Fisheries, Victoria BC, Canada.

Peters, G. & Quang Hong, L. (1984). The effect of social stress on gill structure and plasma electrolyte levels of European eels (*Anguilla anguilla* L.). *Verfahren Deutsch Zoologisch Gesellschaft*, **7**, 318–22.

Peters, G., Faisal, M., Lang, T. & Ahmed, I. (1988). Stress caused by social interaction and its effect on susceptibility to *Aeromonas hydrophila* infection in rainbow trout *Salmo gairdneri*. *Diseases of Aquatic Organisms*, **4**, 83–89.

Pickering, A.D. (1984). Cortisol-induced lymphocytopenia in brown trout, *Salmo trutta* L. *General and Comparative Endocrinology*, **53**, 252–9.

Pickering, A.D. & Stewart, A. (1984). Acclimation of the interrenal tissue of the brown trout, *Salmo trutta* L., to chronic crowding stress. *Journal of Fish Biology*, **24**, 731–40.

Piper, R.G., McElwain, I.B., Orme, L.E., McCraren, J.P., Fowler, L.G. & Leonard, J.R. (1982). Transportation of live fishes. In *Fish Hatchery Management*. US Fish and Wildlife Service, Washington DC.

Plumb, J.A. (1994). *Health Maintenance of Cultured Fishes: Principal Microbial Diseases.* CRC Press, Boca Raton, Florida.

Post, G. (1987). *Textbook of Fish Health.* TFH Publications, Neptune City, New Jersey.

Ravndal, J., Løvold, T., Bentsen, H.B., Røed, K.K., Gjedrem, T. & Røvik, K. (1994). Serum iron levels in farmed Atlantic salmon: family variation and associations with disease resistance. *Aquaculture*, **125**, 37–45.

Romalde, L., Magariños, B., Pazos, F., Silva, A. & Toranzo, A.E. (1994). Incidence of *Yersinia ruckeri* in two farms in Galicia (NW Spain) during a one-year period. *Journal of Fish Diseases*, **17**, 533–9.

Salonius, K. & Iwama, G.K. (1993). Effects of early rearing environment on stress response, immune function, and disease resistance in juvenile coho (*Oncorhynchus kisutch*) and chinook salmon (*O. tshawytscha*). *Canadian Journal of Fisheries and Aquatic Sciences*, **50**, 759–66.

Smith, C.E. (1978). Transportation of salmonid fishes. In *Manual of Fish Culture, Section G: Fish Transportation.* pp. 9–41. US Fish and Wildlife Service, Washington DC.

Smith, R.R. (1977). *Studies on the Energy Metabolism of Cultured Fishes.* Doctoral Dissertation, Graduate School of Cornell University, Cortland, New York.

Snieszko, S.F. (1954). Therapy of bacterial fish diseases. *Transactions of the American Fisheries Society*, **83**, 313–30.

Soares, F., Dinis, M.T. & Pousao-Ferreira, P. (1994). Development of the swim bladder of cultured *Sparus aurata* L.: a histological study. *Aquaculture and Fisheries Management*, **25**, 849–54.

Soderberg, R.W., Meade, J.W. & Redell, L.A. (1993). Growth, survival, and food conversion of Atlantic salmon reared at four different densities with common water quality. *Progressive Fish-Culturist*, **55**, 29–31.

Summerfelt, R.C. (1991). *Symposium on Fish and Crustacean Larviculture.* Special Publication no. 15. European Aquaculture Society. Gent, Belgium.

Summerfelt, R.C. & Smith, L.S. (1990). Anesthesia, surgery, and related techniques. In *Methods for Fish Biology.* Schreck, C. & Moyle, P. (eds.) pp. 213–72. American Fisheries Society, Bethesda, Maryland.

Swain, D.P. & Riddle, B.E. (1990). Variation in agonistic behavior between newly emerged juveniles from hatchery and wild populations of coho salmon, *Oncorhynchus kisutch. Canadian Journal of Fisheries and Aquatic Science*, **47**, 566–71.

Thompson, I., White, A., Fletcher, T.C., Houlihan, D.F. & Secombes, C.J. (1993). The effect of stress on the immune response of Atlantic salmon (*Salmo salar*) fed diets containing different amounts of vitamin C. *Aquaculture*, **114**, 1–18.

Tucker, C.S. & Robinson, E.H. (1990). *Channel Catfish Farming Handbook*. Van Nostrand Reinhold, New York.

Wald, L. & Wilzbach, M.A. (1992). Interactions between native brook trout and hatchery brown trout: effects on habitat use, feeding, and growth. *Transactions of the American Fisheries Society*, **121**, 287–96.

Wallace, J.C. & Kolbeinshavn, A.G. (1988). The effect of size grading and subsequent growth in fingerling Arctic charr, *Salvelinus alpinus* (L.). *Aquaculture*, **73**, 97–100.

Walters, G.R. & Plumb, J.A. (1980). Environmental stress and bacterial infections in channel catfish, *Ictalurus punctatus* Rafinesque. *Journal of Fish Biology*, **17**, 177–85.

Watten, B.J. (1992). Modeling the effects of sequential rearing on the potential production of controlled environment fish-culture systems. *Aquacultural Engineering*, **11**, 33–46.

Wedemeyer, G. (1996*a*). *Physiology of Fish in Intensive Culture*. Chapman & Hall, New York.

Wedemeyer, G.A. (1996*b*). Handling and transportation of salmonids. In *Principles of Salmonid Aquaculture*. Pennel, W. & Barton, B. (eds.) Elsevier Publishing, Netherlands, in press.

Wedemeyer, G.A. & Goodyear, C.P. (1984). Diseases caused by environmental stressors. In *Diseases of Marine Animals*, Volume 4. Part 1: *Pisces*. Kinne, O. (ed.) pp. 424–34. Biologische Anstalt Helgoland, Hamburg.

Wedemeyer, G.A., Meyer, F.P. & Smith, L. (1976). *Environmental Stress and Fish Diseases*. TFH Publications, Neptune City, New Jersey.

Westers, H. (1984). *Principles of Intensive Fish Culture (A Manual for Michigan's State Fish Hatcheries)*. Michigan Department of Natural Resources, Lansing, Michigan.

Wise, D.J., Schwedler, T.E. & Otis, D.L. (1993). Effects of stress on susceptibility of naive channel catfish in immersion challenge with *Edwardsiella ictaluri*. *Journal of Aquatic Animal Health*, **5**, 92–7.

N.W. PANKHURST and G. VAN DER KRAAK

Effects of stress on reproduction and growth of fish

Introduction

It is widely recognized that practices common in aquaculture today, ranging from capture of wild fish for the collection of gametes, social interactions at non-natural stocking densities through to routine husbandry procedures such as handling and confinement, are stressful to fish and may affect reproduction and growth. Stress has inhibitory effects on reproduction in every species in which the relationship has been examined. Stress effects can be manifest at various levels of the reproductive endocrine axis, on gamete development and quality, as well as on subsequent egg and larval survival and development. Stress inhibits growth by exerting metabolic effects and by affecting the endocrine pathways regulating growth. The regulation of growth in fishes, as in other vertebrates, is complex and not as well understood as the control of reproduction. Unlike reproduction, no single endocrine cascade is responsible for the control of growth, making the assessment of the specific effects of stress on growth difficult.

This chapter describes the manner by which stress influences the endocrine pathways controlling reproduction and growth physiology in fish with specific emphasis on the effects of stress and its associated endocrine changes on gamete quality and energy utilization. The final section deals with an evaluation of the consequences of stress on reproductive and growth performance from the perspective of the long term management of wild and domesticated populations of fish.

Stress and reproduction

Endocrine effects

Endocrine regulation of reproduction is achieved via the dual actions of gonadotropin-releasing hormone (GnRH; stimulatory) and dopamine (DA; inhibitory) released from hypothalamic neurons that synapse directly with gonadotropes in the pars distalis of the pituitary (Fig. 1).

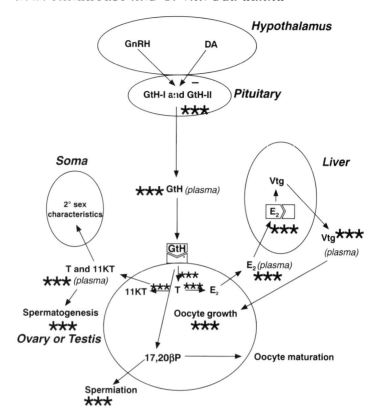

Fig. 1. Summary of the reproductive endocrine system of teleost fishes showing the levels at which stress has been demonstrated to have an inhibitory effect (asterisks). Pathways are stimulatory unless shown otherwise (−). Feedback loops are not shown. Abbreviations: DA – dopamine, E_2 – 17β-estradiol, GnRH – gonadotropin releasing hormone, GtH – gonadotropin, 11KT – 11-ketotestosterone, 17,20βP – 17α,20β-dihydroxy-4-pregnen-3-one, T – testosterone, Vtg – vitellogenin.

Gonadotropes release two gonadotropins (GtH-I and GtH-II) into the systemic circulation for transport to membrane-bound receptors on cells in the ovary and testis. GtH-I and GtH-II share the same spectrum of biological activities in terms of stimulatory effects on ovarian DNA synthesis, ovarian and testicular steroid biosynthesis and induction of oocyte final maturation (Swanson, 1991; Van Der Kraak *et al.* 1992*a*; Srivastava & Van Der Kraak, 1994). The gonads, in turn, produce three main classes of steroids: progestins, androgens, and estrogens. Progestins are C_{21} steroids primarily associated with final oocyte maturation in females and spermiation in males. Two of the most important progestins in fish are 17α,20β-dihydroxy-4-pregnen-3-one [17,20βP] and 17α,20β,21-trihydroxy-4-pregnen-3-one [17,20β,21P]. Androgens are responsible for regulation of testis growth and secondary sex characteristics in males, central feedback in both sexes, and act as precursors for estrogen production. The main androgens produced by fish are testosterone (T) and 11-ketotestosterone (11KT). Estrogens [mainly 17β-estradiol (E_2)] stimulate the hepatic synthesis of the yolk precursor vitellogenin (Vtg), which is then taken up into the growing oocytes under the influence of GtHs (reviewed in Scott & Canario, 1987; Pankhurst & Carragher, 1991; Peter *et al.*, 1991; Van Der Kraak *et al.*, 1992a; Thomas, 1994).

There is currently no information on the effects of stress on the release of GnRH or DA, or on their action on the gonadotropic cells in the pituitary. However, stress is known to result in changes in plasma GtH levels. Acute confinement stress in male brown trout *Salmo trutta* resulted in short-term increases in plasma GtH (Pickering *et al.*, 1987; Sumpter *et al.*, 1987), whereas capture and transport of white sucker *Catostomus commersoni* resulted in depression of GtH to undetectable levels within 24 h of capture (Stacey *et al.*, 1984). At this stage, it is not possible to determine whether the differences between trout and suckers represent species-specific responses, differences generated by the type of stress involved in each case, or whether they arise from the use of domesticated (trout) versus wild fish (suckers). In general terms, it appears that stress will affect plasma GtH levels, but the regulation and nature of such changes remain largely undescribed.

The effect of stress on plasma levels of gonadal steroids is less ambiguous, with acute or chronic stress resulting in depression of plasma levels of androgens and/or estrogens in most species. Male brown trout exposed to acute handling and confinement for 1 h had depressed plasma levels of T and 11KT and a large increase in plasma levels of cortisol within 1 h and 4 h, respectively, of exposure to the stress (Pickering *et al.*, 1987; Sumpter *et al.*, 1987). A month of confinement (chronic stress) also elevated plasma cortisol and depressed plasma T and 11KT levels (Pickering *et al.*, 1987). Female brown and male rainbow trout *Oncorhynchus mykiss* exposed to 2 weeks of confinement stress (with subsequent elevation of plasma cortisol) had reduced plasma levels of T compared to unstressed fish, whereas female rainbow and male brown trout showed no changes in plasma steroid levels. Plasma levels of E_2 were unaffected by confinement in either female brown or rainbow trout (Campbell, Pottinger & Sumpter, 1994). Wild female rainbow trout had reduced plasma levels of T and E_2 and markedly increased plasma cortisol levels 24 h after line capture from a natural spawning run and subsequent confinement in cages within the spawning stream (Pankhurst & Dedual, 1994).

Stress also depresses plasma levels of gonadal steroids in non-salmonids. Line-caught snapper *Pagrus auratus* had reduced plasma E_2 and T levels 1 h and 6 h, respectively, after capture. Plasma T and E_2 levels remained at low levels for the following 5 days (Carragher & Pankhurst, 1991). Wild female red gurnard *Chelidonichthys kumu* showed a similar depression of plasma T and E_2 levels following line capture and return to the laboratory (Clearwater, 1992). Line capture of wild male spotted sea trout *Cynoscion nebulosis* resulted in increased plasma cortisol levels in 30 min and a fall in plasma T levels during the first 60 min after capture. Female spotted sea trout captured in the same way and returned to the laboratory still had an elevated plasma cortisol and low levels of T and E_2 1 day after capture (Safford & Thomas, 1987). White sucker captured in trap nets showed a marked fall in T in preovulatory females and a similar fall in 11KT and T in sexually mature males 24 h after capture and handling (Van Der Kraak *et al.*, 1992b). Subsequent studies with white sucker showed that a 5-min exposure to air reduced circulating steroid levels in both sexes within 1 h and that steroid levels did not recover following a 1-day or 3-day holding period (McMaster *et al.*, 1994; Jardine, Van Der Kraak & Munkittrick, 1996).

Not all gonadal steroids decrease in response to stress. Plasma levels of the maturational steroid 17,20βP were elevated in stressed female snapper (Carragher & Pankhurst, 1991). In contrast, plasma 17,20βP levels were not affected by the stressors that resulted in depression of plasma levels of T and E_2 in wild rainbow trout (Pankhurst & Dedual, 1994). Similar results were found in relation to the white sucker, for which capture and handling had no effect on plasma 17,20βP levels (Van Der Kraak *et al.*, 1992b; McMaster *et al.*, 1994). This suggests that some steroid-converting enzymes (for example, $P450_{C17}$ mediating the conversion of progestins to androgens) are more sensitive to stress than others, perhaps providing the opportunity for unaffected pathways (for example, 20β-hydroxysteroid-dehydrogenase-mediated conversion of 17α-hydroxyprogesterone to 17,20βP) to operate from a larger substrate pool. The concept that $P450_{C17}$ is affected by stress in fish (Van Der Kraak *et al.*, 1992b) is consistent with studies of the rat showing that acute immobilization stress inhibits both 17α-hydroxylase and

17,20-lyase activities which are catalyzed by this microsomal P450 enzyme (Orr *et al.*, 1994). A similar effect was seen in African catfish *Clarias gariepinus*, where *in vitro* incubation of testes from wild fish resulted in reduced production of androgens and increased production of C_{21} steroids, compared to incubations of testes from domesticated fish. The results were interpreted as arising from the greater sensitivity to stress of the wild fish (Schoonen & Lambert, 1986).

Gamete quality

Stress-induced depression of plasma levels of gonadal steroids appears to be associated with a higher incidence of ovarian atresia. In both snapper and red gurnard, confinement of wild fish results in atresia of the majority of vitellogenic follicles within a few days (J. Cleary & N.W. Pankhurst, unpublished data; Clearwater, 1992). Similarly, the onset of gonadal atresia was observed to occur in northern pike *Esox lucius* soon after capture (De Montalembert, Jalabert & Bry, 1978). This suggests that, in some cases, the effects of stress may be irreversible in terms of the current reproductive season. By comparison, mature post-vitellogenic follicles appear to be unaffected by stress and will continue to undergo final maturation and ovulation either spontaneously or in response to exogenous hormone treatment (Carragher & Pankhurst, 1991; Pankhurst & Carragher, 1992).

Stress can also have more subtle effects on gamete quality. Repeated episodes of acute stress (3 min of emersion) randomly applied to rainbow trout over the 9 months prior to spawning resulted in delayed ovulation and reduced egg size in females and lowered sperm counts in males. Progeny from stressed fish had lower survival up to 28 days post-hatch (Campbell, Pottinger & Sumpter, 1992). Chronic stress (2 weeks of confinement) resulted in a fall in plasma Vtg levels and production of smaller eggs in female rainbow trout whereas sperm counts of males were unaffected. Survival of the progeny of stressed rainbow and brown trout was lower than that of progeny of unstressed fish (Campbell *et al.*, 1994). Both of these studies suggest that reduced egg size alone does not account for the lower survival of progeny of stressed fish; at this stage, the source of the effect is unknown.

Stress also appears to affect gamete quality in non-salmonids. Atlantic cod *Gadus morhua* subjected to netting stress three times a week exhibited a higher incidence of deformed larvae, even though there were no differences in egg production, fertilization and survival between progeny of stressed and unstressed fish (Short, Crim & Morgan, 1995).

Mediation of stress effects

As one of the consistent primary results of stress is an increase in circulating levels of cortisol, it has been generally assumed that cortisol mediates the effects of stress on reproduction. This has been investigated in studies where fish have been treated with exogenous cortisol. *In vivo* treatment of rainbow and brown trout with cocoa butter pellets containing cortisol was found to elevate plasma cortisol and depress gonad size and plasma T levels in male brown trout, to depress gonad size, plasma T, E_2 and Vtg levels in female brown trout, to depress plasma GtH in male rainbow trout and to depress plasma Vtg levels in immature female rainbow trout (Carragher *et al.*, 1989). Fish treated with cortisol also had a high incidence of opportunistic infection and mortality, making it difficult to assess whether cortisol has a primary effect on reproduction, or an indirect effect due to the poor health of the fish (presumably a result of the immunosuppressive effects of cortisol, see chapter by P.H.M. Balm, this volume). Evidence for a direct effect of cortisol was provided in a later study of mature male rainbow trout, in which *in vitro* incubation of pituitaries with 1000 ng ml^{-1} cortisol resulted in a lower rate of release of GtH into the medium and a lower final amount of GtH in the pituitary compared with pituitaries treated with cortisol at 0, 10 or 100 ng ml^{-1} (Carragher & Sumpter, 1990*a*). In another study, implantation with cortisol pellets depressed E_2 and Vtg levels in sexually mature female brown trout and depressed Vtg (but not E_2) in immature females (Pottinger, Campbell & Sumpter, 1991). Receptor binding of E_2 by the hepatic cytosol was also impaired, suggesting that the effects of cortisol may arise from interference of E_2 stimulation of hepatic synthesis of Vtg. An effect of this type could account for the situations described above where stress impaired gamete quality but did not appear to suppress plasma levels of E_2.

Implantation of tilapia *Oreochromis mossambicus* with pellets containing cortisol also elevated plasma cortisol levels to levels typical of those found in acutely stressed fish. This was associated with decreased plasma levels of T and E_2, smaller oocyte diameter and gonad size in females, and reduced plasma T levels in males (Foo & Lam, 1993*a,b*). Here also, long-term elevation of plasma cortisol appeared to have somatic effects, with all treated fish losing body mass. Under such conditions, it is possible that energy is diverted from reproductive processes to meet other metabolic requirements. There seems to be no question that exogenous cortisol can impair reproductive function, but it is still not clear whether this effect is direct, or indirect via cortisol-induced changes in metabolism or immunocompetence.

Several workers have examined the direct effects of cortisol on ovarian steroidogenesis using *in vitro* preparations. Rainbow trout ovarian follicles had depressed basal T and E_2 production after incubation with cortisol. This effect was found in seven out of eight fish tested, but sensitivity varied, with responses being observed at cortisol concentrations as low as 5 ng ml^{-1} or as high as 1000 ng ml^{-1} (Carragher & Sumpter, 1990b). Surprisingly, cortisol had no effect on GtH-stimulated steroidogenesis. This was in contrast to an earlier pilot study in which cortisol, at 100 ng ml^{-1}, suppressed both basal and GtH-stimulated production of E_2 (Sumpter *et al.*, 1987). More recently, a repeat of this earlier work, using a similar protocol, failed to find, consistently, an inhibitory effect of cortisol on E_2 production by rainbow trout ovarian follicles (Pankhurst, Van Der Kraak & Peter, 1995a). A small depression in basal E_2 production was found in two fish, but cortisol had no effect on follicles from a further six fish.

Studies of non-salmonid fishes provide no evidence for direct suppression of T or E_2 production by cortisol. *In vitro* incubation of ovarian follicles of goldfish *Carassius auratus*, common carp *Cyprinus carpio* and snapper with cortisol at doses of 1–1000 ng ml^{-1} had no effect on the basal production of T and E_2, or on that stimulated by GtH or the steroid precursor 25-hydroxycholesterol (Pankhurst, Van Der Kraak & Peter, 1995b). The lack of a consistent effect of cortisol on ovarian steroidogenesis in trout and the absence of any effect in other species suggests that the effects of stress on reproduction are not expressed through the effects of cortisol at the ovarian level. If this is the case, then the possibilities are that cortisol acts elsewhere in the endocrine cascade, or that it exerts its effects on reproduction by indirect mechanisms, or that some other stress factors are involved in the suppression of reproduction.

The possible actions of cortisol elsewhere in the reproductive endocrine pathway remain largely unexplored, and there is also very little information on the effects of other stress factors on reproduction. In addition to increases in plasma cortisol levels, stress results in short-term increases in plasma catecholamines (reviewed in Barton & Iwama, 1991) and more sustained increases in plasma levels of adrenocorticotrophic hormone (ACTH), α-melanocyte stimulating hormone (α-MSH) and endorphins (Kawauchi, 1983; Sumpter, 1986; Sumpter, Dye & Benfey, 1986). ACTH tested at a dose of 100 ng ml^{-1} had no effect on basal or GtH-stimulated E_2 production by rainbow trout ovarian follicles (Sumpter *et al.*, 1987) and preliminary studies indicate that α-MSH and β-endorphin also have no *in vitro* effect on E_2 production by rainbow trout follicles (N.W. Pankhurst, unpublished data). However, this work should be repeated and the possible direct effects of other stress hormones (e.g., cytokines) on ovarian steroidogenesis should be considered.

There is also conflicting evidence as to whether cortisol is directly responsible for the lower quality of eggs produced by stressed fish. Maternal steroids, including cortisol, are sequestered by growing oocytes and are thought to influence the developmental profile of the embryos (Hwang et al., 1992; de Jesus, Hirano & Inui, 1993). This suggests that stressed fish have eggs with a higher cortisol content and that this will adversely affect rates or patterns of embryonic development. There is little evidence to show that eggs of stressed fish do in fact have higher cortisol contents than those of unstressed fish. Moreover, treatment of rainbow trout with cortisol pellets, which did elevate the cortisol content of the eggs, did not have any effect on survival to hatching (Brooks et al., 1995). However, this study did not investigate whether subsequent larval survival or development was impaired.

Stress may also interfere with reproduction by affecting other endocrine processes (see chapter by J.P. Sumpter, this volume). Pottinger, Prunet and Pickering (1992) reported that 48 h of confinement stress increased plasma cortisol levels with a concomitant depression of plasma prolactin (PRL) levels. PRL has a central role in osmoregulation and changes in PRL levels are, therefore, likely to lead to osmotic imbalances. These may in turn interfere with ion-dependent processes in reproduction, such as the changes in intracellular Ca^{2+}, Na^+, or Cl^- content which are involved in the signal transduction pathways for GtH-stimulated ovarian steroidogenesis (Van Der Kraak, 1991, 1992, unpublished). There is also evidence that environmental stress, in the form of pollution, has an inhibitory effect on reproduction. Salmonids respond to environmental acid stress with lowered plasma E_2, Vtg and delayed ovulation (Tam & Payson, 1986; Weiner, Schreck & Li, 1986; Mount et al., 1988; Roy et al., 1990; Tam et al., 1990). Exposure to pulp mill effluent has similar effects on white suckers, with exposed fish having an increased age to sexual maturity, lower fecundity and egg size and decreases in plasma levels of GtH and gonadal steroids compared to fish from unpolluted sites (McMaster et al., 1991; Munkittrick et al., 1991; Van Der Kraak et al., 1992b). It is not clear whether these effects are the direct result of the exposure to pollutants, or whether they arise from the secondary effects of stress caused by that exposure, but it seems likely that both might occur (see also Kime, 1995). Assessment of the effects of pollutants in terms of generating a stress response which in turn inhibits reproduction is complicated by changes in water quality and pollutants which may modulate the corticosteroid response to subsequent stress (Pickering & Pottinger, 1987; Hontela *et al.*, 1992; McMaster *et al.*, 1994). Nevertheless, it is clear that environmental stress in the broader sense appears to have the same capacity to inhibit reproduction as acute or chronic stresses associated with capture and handling procedures.

Stress and growth

Endocrine effects

The multiple endocrine systems participating in the regulation of growth and metabolic processes in fish are summarized in Fig. 2 (see also Sumpter, 1992 for a review of control of growth processes in fish). Growth hormone (GH) is released from the pituitary, probably as in mammals, under the dual control of a releasing hormone (excitatory) and somatostatin (inhibitory). Both GnRH and DA have been shown to stimulate the release of GH from the teleost pituitary (Marchant *et al.*, 1989; Chang, Jobin & Wong, 1993; Wong, Chang & Peter, 1993). GH stimulates the production and release of insulin-like growth factor (IGF-1) from the peripheral tissues (principally the liver), where it can either have paracrine action or be carried in the circulation, attached to specific carrier proteins, to somatic tissue receptors. IGF-1 affects cell growth and differentiation. Growth processes are further modified by the actions of other endocrine factors. Insulin produced by the pancreatic islets has growth-promoting effects, both via its partial cross reaction with IGF-1 receptors and also by optimizing IGF-1 production by modulating the affinity of hepatic GH receptors. Insulin also promotes growth through its actions in stimulating glycogen storage, lipogenesis and the uptake and incorporation of amino acids into proteins (see Plisetskaya, Duguay & Duan, 1994 for an overview of insulin and IGF-1 in salmonid fish). The thyroid hormones thyroxine (T_4) and its peripheral metabolite triiodothyronine (T_3) are also required for normal growth (Leatherland, 1994), and are thought to have permissive effects on both GH secretion and on the cellular response to IGF-1. Gonadal steroids, particularly androgens, can have powerful growth-promoting effects (Fig. 2).

Stress results in variable changes in plasma levels of GH. Pickering *et al.* (1991) found that acute stress depressed plasma GH levels of rainbow trout, whereas chronic stress resulted in elevation of GH. On the basis that plasma GH levels also increased in starved fish (Sumpter *et al.*, 1991), Pickering *et al.* (1991) hypothesized that the apparently paradoxical effects of chronic stress could have arisen from starvation

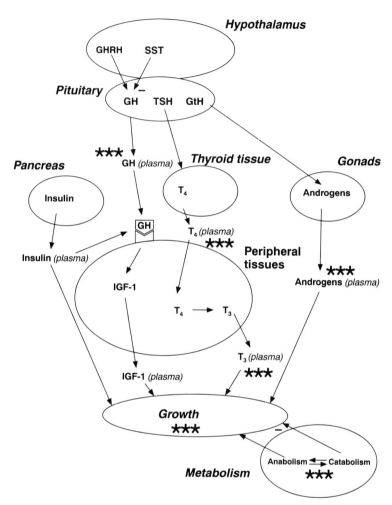

Fig. 2. Summary of the endocrine and metabolic process controlling growth in teleost fishes, showing the levels at which stress has been demonstrated to have an inhibitory effect (asterisks). Pathways are stimulatory unless shown otherwise (−). Feedback loops are not shown. Abbreviations: GH – growth hormone, GHRH – growth hormone releasing hormones, GtH – gonadotropin, IGF-1 – insulin-like growth factor-1, SST – somatostatin, T_3 – triiodothyronine, T_4 – thyroxine, TSH – thyroid stimulating hormone.

effects (stressed fish ate less). Similar results were reported by Farbridge and Leatherland (1992) where fasted rainbow trout had higher GH levels than fed fish, but both groups showed a depression of plasma GH in response to acute stress. In contrast, Cook and Peter (1984) found that acute stress resulted in increases in plasma GH in goldfish, and Wagner and McKeown (1986) found that stress had no effect on plasma GH in salmon. There is also evidence for stimulatory effects of cortisol on *in vitro* GH release from pituitary glands of tilapia (Nishioka, Grau & Bern, 1985) and a potentiating effect of GH on cortisol release by the interrenal tissue of coho salmon *Oncorhynchus kisutch* (Young, 1988). This appears to be consistent with the synergistic effect of cortisol and GH in seawater adaption of salmonids (Madsen, 1990; Madsen & Korsgaard, 1991). It is clear that there is an interaction between stress and GH and, probably, also between cortisol and GH, but the nature of the interaction requires further study. There is also limited information on the effects of stress and/or cortisol on GH stimulation of IGF-1 production or the tissue actions of IGF-1 (Plisetskaya *et al.*, 1994).

Stress may also exert growth-inhibiting effects through depression of T_3 and T_4 levels, either directly or indirectly by changes in the nutritional state arising from a stress-induced reduction in food intake (reviewed in Pickering, 1993). Here there is better evidence for stress effects being mediated by cortisol. Cortisol treatment was found to suppress T_3 and T_4 levels in the eel *Anguilla anguilla* (Redding, de Luze & Leloup-Hatey, 1986) and T_3 levels in a number of salmonids (Redding *et al.*, 1984; Vijayan & Leatherland, 1989; Brown *et al.*, 1991). The effect appears to be generated by an increase in the plasma clearance rate of T_3 (Brown *et al.*, 1991). Finally, the observed anabolic effects of androgens (reviewed in Sumpter, 1992) are likely to be suppressed by the inhibitory effects of stress on reproduction, as described in the previous section of this chapter (p. 75).

Metabolic consequences of stress

Stress results in an increase in energy demand, met mainly from the mobilization or synthesis of glucose from stored reserves. Under natural conditions, this increase in available energy is used to meet and resolve the challenge imposed by the stressor. However, fish in captivity may be exposed to stress for extended periods without the scope to resolve the stressor, with the result that metabolism also remains geared up for extended periods. Diversion of energy under these circumstances reduces the energy available for other requirements, including somatic growth. This effect is demonstrated by the increase in oxygen consumption shown by stressed fish and the elevated plasma glucose levels consistently associated with stress (reviewed in Barton & Iwama, 1991).

The short-term increases in catecholamines occurring in response to acute stress are known to stimulate increases in plasma glucose via the mobilization of glycogen stores. However, in view of the relatively transitory elevations of plasma catecholamines in response to stress (Barton & Iwama, 1991), it is unlikely that catecholamine-mediated metabolism is responsible for chronically elevated plasma glucose levels in stressed fish. There is some evidence that cortisol is responsible for the observed increases in plasma glucose levels. Cortisol may increase the free amino acid pool, either by inhibiting protein synthesis or by stimulating protein catabolism via the enhanced activity of hepatic aminotransferases. This may make free amino acids available for gluconeogenesis (reviewed in van der Boon, van den Thillart & Addink, 1991), although there is limited evidence to show that this actually occurs. Treatment of fish with cortisol consistently depresses somatic growth and this is often associated with hyperglycemia (Butler, 1968; Chan & Woo, 1978; Lidman et al., 1979; Leach & Taylor, 1982; Davis et al., 1985; Barton, Schreck & Barton, 1987; Vijayan & Leatherland, 1989; Foo & Lam, 1993a,b). The best evidence that this effect is generated directly by cortisol comes from work by Leach and Taylor (1980), in which stress-induced hyperglycemia was reduced in mummichog Fundulus heteroclitus treated with the 11β-hydroxylase inhibitor, metyrapone (inhibiting the synthesis of cortisol). Alternatively, increased plasma glucose concentrations may result from peripheral inhibition of glucose utilization (Leach & Taylor, 1982). The conclusions of van der Boon et al. (1991), i.e., that cortisol had metabolic effects but that the nature of these was variable and that the evidence was based on a very limited number of studies, still apply at the present time.

Management considerations

Stress-induced inhibition of growth and reproduction has severe consequences for management of both wild and domestic populations of fish. As previously discussed, detrimental changes in environmental quality (from whatever source) will have the potential to reduce the growth and reproductive fitness of fish populations inhabiting those waters, with consequent effects on population size and structure. Short-term imposition of acute stress via exposure to fishing gear may also have the capacity to interfere with natural reproduction. Capture by line or trawl is stressful (Pankhurst & Sharples, 1992) and the laboratory experiments described earlier show that this type of stress can have lasting effects on reproduction. If these effects also occur in fish captured and then subsequently released back to the wild (for example via escapement of undersized fish through net meshes, or live return of undersized fish from nets or lines), then the impact of fishing on natural populations may be greater than fishing mortality alone. The same considerations apply in sport fisheries, particularly those targeting salmonids, where there is increasing reliance on catch-and-release management strategies. Such strategies need to be evaluated in terms of the possible effects of capture and handling on subsequent reproductive success.

The problems are more immediate in aquaculture. Use of wild broodstock is the mainstay of most developing aquaculture ventures, particularly those using marine species. Unless techniques can be developed for ameliorating stress effects imposed by capture and handling, egg production will continue to be constrained. A partial solution lies in the development of domestic broodstock, with anecdotal evidence suggesting that stress effects on reproduction decline with each cultured generation. This is supported by data showing that the stress responses of wild and domestic fish differ (Woodward & Strange, 1987) and that wild fish do not appear to acclimate to captivity in terms of their response to stress (Salonius & Iwama, 1993). However, the demonstrated effects of stress on egg and larval quality, even when apparently normal gamete development occurs, suggest that stress will continue to be a problem in domesticated stocks. An effect of this type may explain the difficulties that many workers have experienced in rearing the typically small larvae of marine species. These difficulties are generally attributed to inappropriate larval rearing protocols, but may well also arise from stress effects on broodstock, ultimately impairing the fitness of progeny.

In addition to contributing to the successful husbandry of fish in culture, solution of the stress–reproduction conundrum has important implications for stock enhancement. Increasingly, marine aquaculture projects have, as part of their focus, enhancement of depleted natural stocks by release to the wild of hatchery-reared juveniles. Research using salmonids has shown that there is a heritable basis for the severity of the stress response (Fevolden, Refstie & Roed, 1991; Pottinger, Moran & Morgan, 1994; see also chapter by T.G. Pottinger and A.D. Pickering, this volume) with the result that domestication may push fish through a narrow selection gate in which stress-sensitive individuals are rapidly lost from the population. If sensitivity to stress is part of a suite of characteristics that enhances survival in the wild, then for enhancement projects to be successful it will probably be necessary to restock with fish that have characteristics that do not make them very amenable to rearing in culture.

The economic viability of aquaculture rests firmly with the growth rates and food conversion efficiencies shown by stock. While the mechanisms by which stress inhibits growth are still far from understood, it is clear that stress-evoked suppression of growth and reduction of food conversion efficiency are potentially major problems in aquaculture. The size of these problems is still not clear, largely because of the lack of perception of stress as a management priority at the farm level. In the same way that broodstock management of stressed fish will remain problematic, the real growth potential of existing or new species for culture may remain unrealized if stress reduction is not a priority of husbandry.

References

Barton, B.A. & Iwama, G.K. (1991). Physiological changes in fish from stress in aquaculture with emphasis on the response and effects of corticosteroids. *Annual Review of Fish Diseases*, **1**, 3–26.

Barton, B.A., Schreck, C.B. & Barton, L.D. (1987). Effects of chronic cortisol administration and daily acute stress on growth, physiological conditions, and stress responses in juvenile rainbow trout. *Diseases of Aquatic Organisms*, **2**, 173–185.

Brooks, S., Pottinger, T.G., Tyler, C.R. & Sumpter, J.P. (1995) Does cortisol influence egg quality in the rainbow trout *Oncorhynchus mykiss*? In *Reproductive Physiology of Fish 1995*. Goetz, F.W. & Thomas, P. (eds.) p. 180. University of Texas, Austin.

Brown, S.B., MacLatchy, D.L., Hara, T.J. & Eales, J.G. (1991). Effect of cortisol on aspects of 3,5,3'-triiodo-L-thyronine metabolism in rainbow trout (*Oncorhynchus mykiss*). *General and Comparative Endocrinology*, **81**, 207–16.

Butler, D.G. (1968). Hormonal control of gluconeogenesis in the North American eel (*Anguilla rostrata*). *General and Comparative Endocrinology*, **10**, 85–91.

Campbell, P.M., Pottinger, T.G. & Sumpter, J.P. (1992). Stress reduces the quality of gametes produced by rainbow trout. *Biology of Reproduction*, **47**, 1140–50.

Campbell, P.M., Pottinger, T.G. & Sumpter, J.P. (1994). Preliminary evidence that chronic confinement stress reduces the quality of gametes produced by brown and rainbow trout. *Aquaculture*, **120**, 151–69.

Carragher, J.F. & Pankhurst, N.W. (1991). Stress and reproduction in a commercially important marine fish, *Pagrus auratus* (Sparidae).

In *Proceedings of the Fourth International Symposium on Reproductive Physiology of Fish*, University of East Anglia, UK, 7–12 July 1991. Scott, A.P., Sumpter, J.P., Kime, D.E. & Rolfe, M.S. (eds.) pp. 253–5. FishSymp 91, Sheffield.

Carragher, J.F. & Sumpter, J.P. (1990*a*). Corticosteroid physiology in fish. In *Progress in Comparative Endocrinology*. Epple, A., Scanes, C.G. & Stetson, M.H. (eds.) pp. 487–492. Wiley-Liss, New York.

Carragher, J.F. & Sumpter, J.P. (1990*b*). The effect of cortisol on the secretion of sex steroids from cultured ovarian follicles of rainbow trout. *General and Comparative Endocrinology*, **77**, 403–7.

Carragher, J.F., Sumpter, J.P., Pottinger, T.G. & Pickering, A.D. (1989). The deleterious effects of cortisol implantation on reproductive function in two species of trout, *Salmo trutta* L. and *Salmo gairdneri* Richardson. *General and Comparative Endocrinology*, **76**, 310–21.

Chan, D.K.O. & Woo, N.Y.S. (1978). Effect of cortisol on the metabolism of the eel, *Anguilla japonica*. *General and Comparative Endocrinology*, **35**, 205–15.

Chang, J.P., Jobin, R.M. & Wong, A.O.L. (1993). Intracellular mechanisms mediating gonadotropin and growth hormone release in the goldfish, *Carassius auratus*. *Fish Physiology and Biochemistry*, **11**, 25–33.

Clearwater, S.J. (1992). Reproductive biology and response to capture stress of red gurnard *Chelidonichthys kumu* (Family: Triglidae). MSc Thesis, University of Auckland, Auckland. p. 96.

Cook, A.F. & Peter, R.E. (1984). Effects of somatostatin on serum growth hormone levels in the goldfish, *Carassius auratus*. *General and Comparative Endocrinology*, **54**, 109–113.

Davis, K.B., Torrance, P., Parker, N.C. & Suttle, M.A. (1985). Growth, body composition, and hepatic tyrosine aminotransferase activity in cortisol-fed channel catfish, *Ictalurus punctatus* Rafinesque. Journal of Fish Biology, **27**, 177–84.

De Montalembert, G., Jalabert, B. & Bry, C. (1978). Precocious induction of maturation and ovulation in northern pike (*Esox lucius*). *Annales de Biologie Animale Biochimie Biophysique*, **18**, 969–75.

Farbridge, K.J. & Leatherland, J.F. (1992). Plasma growth hormone levels in fed and fasted rainbow trout (*Oncorhynchus mykiss*) are decreased following handling stress. *Fish Physiology and Biochemistry*, **10**, 67–73.

Fevolden, S.E., Refstie, T. & Roed, K.H. (1991). Selection for high and low cortisol stress response in Atlantic salmon (*Salmo salar*) and rainbow trout (*Oncorhynchus mykiss*). *Aquaculture*, **95**, 53–65.

Foo, T.W. & Lam, T.J. (1993*a*). Retardation of ovarian growth and depression of serum steroid levels in the tilapia *Oreochromis mossambicus*, by cortisol implantation. *Aquaculture*, **115**, 133–43.

Foo, T.W. & Lam, T.W. (1993*b*). Serum cortisol response to handling stress and the effect of cortisol implantation on testosterone levels in the tilapia *Oreochromis mossambicus*. *Aquaculture*, **115**, 145–58.

Hontela, A., Rasmussen, J.B., Audet, C. & Chevalier, G. (1992). Impaired cortisol stress response in fish from environments polluted by PAHs, PCBs, and mercury. *Archives of Environmental Contamination and Toxicology*, **22**, 278–83.

Hwang, P-P., Wu, S-W., Lin, J.-H. & Wu, L-S. (1992). Cortisol content of eggs and larvae of teleosts. General and Comparative Endocrinology, **86**, 189–96.

Jardine, J.J., Van Der Kraak, G.J., & Munkittrick, K.R. (1996). Capture and confinement stress in white sucker exposed to bleached kraft pulp mill effluent. *Ecotoxicology and Environmental Safety*, **33**, 287–98.

de Jesus, E.G., Hirano, T. & Inui, Y. (1993). Flounder metamorphosis: its regulation by various hormones. *Fish Physiology and Biochemistry*, **11**, 323–8.

Kawauchi, H. (1983). Chemistry of proopiocortin-related peptides in the salmon pituitary. *Archives of Biochemistry and Biophysics*, **227**, 343–50.

Kime, D.E. (1995). The effects of pollution on reproduction in fish. *Reviews in Fish Biology and Fisheries*, **5**, 52–96.

Leach, G.L. & Taylor, M.H. (1980). The role of cortisol in stress-induced metabolic changes in *Fundulus heteroclitus*. *General and Comparative Endocrinology*, **42**, 219–27.

Leach, G.L. & Taylor, M.H. (1982). The effect of cortisol treatment on carbohydrate and protein metabolism in *Fundulus heteroclitus*. *General and Comparative Endocrinology*, **48**, 76–83.

Leatherland, J.F. (1994). Reflections on the thyroidology of fishes: from molecules to mankind. *Guelph Ichthyology Reviews*, **2**, 1–67.

Lidman, U., Dave, G., Johansson-Sjöbeck, M.L., Larsson, A. & Lewander, K. (1979). Metabolic effects of cortisol in the European eel, *Anguilla anguilla* (L.). *Comparative Biochemistry and Physiology*, **63A**, 339–44.

Madsen, S.S. (1990). The role of cortisol and growth hormone in seawater adaptation and development of hypoosmoregulatory mechanisms in sea trout parr (*Salmo trutta trutta*). *General and Comparative Endocrinology*, **79**, 1–11.

Madsen, S.S. & Korsgaard, B. (1991). Opposite effects of 17β-estradiol and combined growth hormone-cortisol treatment on hypo-osmoregulatory performance in sea trout presmolts, *Salmo trutta*. *General and Comparative Endocrinology*, **83**, 276–82.

Marchant, T.A., Chang, J.P., Nahorniak, C.S. & Peter, R.E. (1989). Evidence that gonadotropin-releasing hormone also functions as a growth hormone releasing factor in the goldfish. *Endocrinology*, **124**, 2509–18.

McMaster, M.E., Van Der Kraak, G.J., Portt, C.B., Munkittrick, K.R., Sibley, P.K., Smith, I.R. & Dixon, D.G. (1991). Changes in hepatic mixed-function oxygenase (MFO) activity, plasma steroid levels and age at maturity of a white sucker (*Catastomus commersoni*) population exposed to bleached kraft pulp mill effluent. *Aquatic Toxicology*, **21**, 199–218.

McMaster, M.E., Munkittrick, K.R., Luxon, P.L. & Van Der Kraak, G.J. (1994). Impact of low level sampling stress on interpretation of physiological responses of white sucker exposed to effluent from a bleached kraft pulp mill. *Ecotoxicology and Environmental Safety* **27**, 251–64.

Mount, D.R., Ingersoll, C.G., Gulley, E.D., Fernandez, J.D., LaPoint, T.W. & Bergman, H.L. (1988). Effect of long term exposure to acid, aluminium, and low calcium on adult brook trout (*Salvelinus fontinalis*). Survival, growth, fecundity, and progeny survival. *Canadian Journal of Fisheries and Aquatic Sciences*, **45**, 1623–32.

Munkittrick, K.R., Portt, C.B., Van Der Kraak, G.J., Smith, I.R. & Rokosh, D. (1991). Impact of bleached kraft mill effluent on population characteristics, liver MFO activity and serum steroid levels of a Lake Superior white sucker (*Catostomus commersoni*) population. *Canadian Journal of Fisheries and Aquatic Sciences*, **48**, 1371–80.

Nishioka, R.S., Grau, E.G. & Bern, H.A. (1985). *In vitro* release of growth hormone from the pituitary gland of tilapia, *Oreochromis mossambicus*. *General and Comparative Endocrinology*, **60**, 90–4.

Orr, T.E., Taylor, M., Bhattacharyya, A.K., Collins, D.C. & Mann, D.R. (1994). Acute immobilization stress disrupts testicular steroidogenesis in adult male rats by inhibiting the activities of 17α-hydroxylase and 17,20-lyase without affecting the binding of LH/hCG receptors. *Journal of Andrology*, **15**, 302–8.

Pankhurst, N.W. & Carragher, J.F. (1991). Seasonal endocrine cycles in marine teleosts. In *Proceedings of the Fourth International Symposium on Reproductive Physiology of Fish 1991*, University of East Anglia, Norwich, UK, 7–12 July 1991. Scott, A.P., Sumpter, J.P., Kime, D.E. & Rolfe, M.S. (eds.) pp. 131–5. FishSymp 91, Sheffield.

Pankhurst, N.W. & Carragher, J.F. (1992). Oocyte maturation and changes in plasma steroid levels in snapper *Pagrus* (= *Chrysophrys*) *auratus* (Sparidae) following treatment with human chorionic gonadotropin. *Aquaculture*, **101**, 337–47.

Pankhurst, N.W. & Dedual, M. (1994). Effects of capture and recovery on plasma levels of cortisol, lactate and gonadal steroids in a natural population of rainbow trout *Oncorhynchus mykiss*. *Journal of Fish Biology*, **45**, 1013–25.

Pankhurst, N.W. & Sharples, D.F. (1992). Effects of capture and confinement on plasma cortisol concentrations in the snapper, *Pagrus auratus*. *Australian Journal of Marine and Freshwater Research*, **43**, 345–56.

Pankhurst, N.W., Van Der Kraak, G. & Peter, R.E. (1995*a*). A reassessment of the inhibitory effects of cortisol on ovarian steroidogenesis. In *Reproductive Physiology of Fish 1995*, Goetz, F.W. & Thomas, P. (eds.) in press. University of Texas, Austin.

Pankhurst, N.W., Van Der Kraak, G. & Peter, R.E. (1995*b*). Evidence that the inhibitory effects of stress on reproduction in teleost fish are not mediated by the action of cortisol on ovarian steroidogenesis. *General and Comparative Endocrinology*, **99**, 249–57.

Peter, R.E., Trudeau, V.L., Sloley, B.D., Peng, C. & Nahorniak, C.S. (1991). Actions of catecholamines, peptides and sex steroids in regulation of gonadotropin-II in the goldfish. In *Reproductive Physiology of Fish 1991*. Scott, A.P., Sumpter, J.P., Kime, D.E. & Rolfe, M.S. (eds.) pp. 30–4. FishSymp 91, Sheffield.

Pickering, A.D. (1993). Growth and stress in fish production. *Aquaculture*, **111**, 51–63.

Pickering, A.D. & Pottinger, T.G. (1987). Poor water quality suppresses the cortisol response of salmonid fish to handling and confinement. *Journal of Fish Biology*, **30**, 363–74.

Pickering, A.D., Pottinger, T.G., Carragher, J. & Sumpter, J.P. (1987). The effects of acute and chronic stress on the levels of reproductive hormones in the plasma of mature male brown trout *Salmo trutta* L. *General and Comparative Endocrinology*, **68**, 249–59.

Pickering, A.D., Pottinger, T.G., Sumpter, J.P., Carragher, J.F. & Le Bail, P.Y. (1991). Effects of acute and chronic stress on the levels of circulating growth hormone in the rainbow trout, *Oncorhynchus mykiss*. *General and Comparative Endocrinology*, **83**, 86–93.

Plisetskaya, E.M., Duguay, S.J. & Duan, C. (1994). Insulin and insulin-like growth factor I in salmonids: comparison of structure, function and expression. In *Perspectives in Comparative Endocrinology*. Davey, K.G., Peter, R.E. & Tobe, S.S. (eds.) pp. 226–233. National Research Council of Canada, Ottawa.

Pottinger, T.G., Campbell, P.M. & Sumpter, J.P. (1991). Stress-induced disruption of the salmonid liver-gonad axis. In *Reproductive Physiology of Fish 1991*. Scott, A.P., Sumpter, J.P., Kime, D.E. & Rolfe, M.S. (eds.) pp. 114–116. FishSymp 91, Sheffield.

Pottinger, T.G., Prunet, P. & Pickering, A.D. (1992). The effects of confinement stress on circulating prolactin levels in rainbow trout (*Oncorhynchus mykiss*) in fresh water. *General and Comparative Endocrinology*, **88**, 454–60.

Pottinger, T.G., Moran, T.A. & Morgan J.A.W. (1994). Primary and secondary indices of stress in the progeny of rainbow trout (*Oncorhynchus mykiss*) selected for high and low responsiveness to stress. *Journal of Fish Biology*, **44**, 149–63.

Redding, J.M., Schreck, C.B., Birks, E.K. & Ewing, R.D. (1984). Cortisol and its effects on plasma thyroid hormone and electrolyte concentrations in fresh water and during sea water acclimation in yearling coho salmon, *Oncorhynchus kisutch*. *General and Comparative Endocrinology*, **56**, 146–58.

Redding, J.M., de Luze, A. & Leloup-Hatey, J. (1986). Suppression of plasma thyroid hormone concentrations by cortisol in European eel, *Anguilla anguilla*. *Comparative Biochemistry and Physiology*, **83A**, 409–13.

Roy, R.L., Ruby, S.M., Idler, D.R. & Ying, S. (1990). Plasma vitellogenin levels in pre-spawning rainbow trout, *Oncorhynchus mykiss*, during acid exposure. *Archives of Environmental Contamination and Toxicology*, **19**, 803–6.

Safford, S.E. & Thomas, P. (1987). Effects of capture and handling on circulating levels of gonadal steroids and cortisol in the spotted seatrout, *Cynoscion nebulosus*. In *Reproductive Physiology of Fish 1987*. Idler, D.R., Crim, L.W. & Walsh, J.M. (eds.) p. 312. Memorial University of Newfoundland, St John's.

Salonius, K. & Iwama, G.K. (1993). Effects of early rearing environment on stress response, immune function, and disease resistance in juvenile coho (*Oncorhynchus kisutch*) and chinook salmon (*O. tshawytscha*). *Canadian Journal of Fisheries and Aquatic Sciences*, **50**, 759–66.

Schoonen, W.G.E.J. & Lambert, J.G.D. (1986). Steroid metabolism in the testes of the African catfish, *Clarias gariepinus* (Burchell), during spawning season under natural conditions and kept in ponds. *General and Comparative Endocrinology*, **61**, 40–52.

Scott, A.P. & Canario, A.V.M. (1987). Status of oocyte maturation-inducing steroids in teleosts. In *Reproductive Physiology of Fish 1987*. Idler, D.R., Crim, L.W. & Walsh, J.M. (eds.) pp. 224–34. Memorial University of Newfoundland, St John's.

Short, C.E., Crim. L.W. & Morgan, M.J. (1995). The effects of stress on spawning performance and larval development in Atlantic cod, *Gadus morhua*. In *Reproductive Physiology of Fish 1995*. Goetz, F.W. & Thomas, P. in press. University of Texas, Austin.

Stacey, N.E., MacKenzie, D.S., Marchant, T.A., Kyle, A.L. & Peter, R.E. (1984). Endocrine changes during natural spawning in the white sucker *Catostomus commersoni*. I. Gonadotropin, growth hormone, and thyroid hormones. *General and Comparative Endocrinology*, **56**, 333–48.

Sumpter, J.P. (1986). ACTH-related material in the pituitary gland of the chinook salmon (*Oncorhynchus tschawytscha*). *General and Comparative Endocrinology*, **62**, 359–66.

Sumpter, J.P. (1992). Control of growth of rainbow trout (*Oncorhynchus mykiss*). *Aquaculture*, **100**, 299–320.

Sumpter, J.P., Carragher, J.F., Pottinger, T.G. & Pickering, A.D. (1987). Interaction of stress and reproduction in trout. In *Reproductive Physiology of Fish 1987*. Idler, D.R., Crim, L.W. & Walsh, J.M. (eds.) pp. 299–302. Memorial University of Newfoundland, St John's.

Sumpter, J.P., LeBail, P.Y., Pickering, A.D., Pottinger, T.G. & Carragher, J.F. (1991). The effect of starvation on growth and plasma growth hormone concentrations of rainbow trout (*Oncorhynchus mykiss*). *General and Comparative Endocrinology*, **83**, 94–102.

Sumpter, J.P., Dye, H.M. & Benfey, T.J. (1986). The effects of stress on plasma ACTH, α-MSH and cortisol levels in salmonid fishes. *General and Comparative Endocrinology*, **62**, 377–85.

Srivastava, R.K. & Van Der Kraak, G. (1994). Regulation of DNA synthesis in goldfish ovarian follicles by hormones and growth factors. *Journal of Experimental Zoology*, **270**: 263–72.

Swanson, P. (1991). Salmon gonadotropins: reconciling old and new ideas. In *Proceedings of the Fourth International Symposium on Reproductive Physiology of Fish 1991*, University of East Anglia, Norwich, UK, 7–12 July 1991. Scott, A.P., Sumpter, J.P., Kime, D.E. & Rolfe, M.S. (eds.) pp. 2–7. FishSymp 91, Sheffield.

Tam. W.H. & Payson, P.D. (1986). Effects of chronic exposure to sublethal pH on growth, egg production, and ovulation in brook trout, *Salvelinus fontinalis*. *Canadian Journal of Fisheries and Aquatic Sciences*, **43**, 275–80.

Tam, W.H., Fryer, J.N., Valentine, B. & Roy, R.L. (1990). Reduction in oocyte production and gonadotrope activity, and plasma levels of estrogens and vitellogenin, in brook trout exposed to low environmental pH. *Canadian Journal of Zoology*, **68**, 2468–79.

Thomas, P. (1994). Hormonal control of final oocyte maturation in sciaenid fishes. In *Perspectives in Comparative Endocrinology*. Davey, K.G., Peter, R.E. & Tobe, S.S. (eds.) pp. 619–25. National Research Council of Canada, Otawa.

van der Boon, J., van den Thillart, G.E.E.J.M. & Addink, A.D.F. (1991). The effects of cortisol administration on intermediary metabolism in teleost fish. *Comparative Biochemistry and Physiology*, **100A**, 47–53.

Van Der Kraak, G. (1991). Role of calcium in the control of steroidogenesis in preovulatory ovarian follicles of the goldfish. *General and Comparative Endocrinology*, **81**, 268–75.

Van Der Kraak, G. (1992). Mechanisms by which calcium ionophore and phorbol ester modulate steroid production by goldfish preovulatory ovarian follicles. *Journal of Experimental Zoology*, **262**, 271–8.

Van Der Kraak, G., Suzuki, K., Peter, R.E., Itoh, H. & Kawauchi, H. (1992*a*). Properties of common carp gonadotropin I and gonadotropin II. *General and Comparative Endocrinology*, **85**, 217–29.

Van Der Kraak, G., Munkittrick, M.E., McMaster, M.E., Portt, C.B. & Chang, J.P.(1992*b*). Exposure to bleached kraft mill effluent disrupts the pituitary-gonadal axis of white sucker at multiple sites. *Toxicology and Applied Pharmacology*, **115**, 224–33.

Vijayan, M.M. & Leatherland, J.F. (1989). Cortisol-induced changes in plasma glucose, protein, and thyroid hormone levels, and liver glycogen content of coho salmon (*Oncorhynchus kisutch* Walbaum). *Canadian Journal of Zoology*, **67**, 2746–50.

Wagner, G.F. & McKeown, B.A. (1986). Development of a salmon growth hormone radioimmunoassay. *General and Comparative Endocrinology*, **62**, 452–8.

Weiner, C.S., Schreck, C.B. & Li, H.W. (1986). Effects of low pH on reproduction of rainbow trout. *Transactions of the American Fisheries Society*, **115**, 75–82.

Wong, A.O.L., Chang, J.P. & Peter, R.E. (1993). Dopamine functions as a growth hormone-releasing factor in the goldfish, *Carassius auratus*. *Fish Physiology and Biochemistry*, **11**, 77–84.

Woodward, C.C. & Strange, R.J. (1987). Physiological stress responses in wild and hatchery-reared rainbow trout. *Transactions of the American Fisheries Society*, **116**, 574–79.

Young, G. (1988). Enhanced response of the interrenal of coho salmon (*Oncorhynchus kisutch*) to ACTH after growth hormone treatment in vivo and in vitro. *General and Comparative Endocrinology*, **71**, 85–92.

J.P. SUMPTER

The endocrinology of stress

Introduction

The endocrine changes associated with the response of fish to stress have received considerable attention during the last 20 years or so. As is often the case, it was a technological breakthrough – in this case, development of radioimmunoassays – that provided the opportunity for the research. Also instrumental in stimulating research in this area was the realization that the consequences of stress to fish were often both marked and deleterious. The development of aquaculture as a major industry also provided a desire to improve our understanding of the physiological changes associated with a stress response, and of the ultimate consequences to the fish of these changes.

This chapter is concerned only with the endocrine changes that have been reported to occur in response to stress; it does not deal with the consequences of these endocrine changes (the so-called secondary and tertiary responses to stress: see Mazeaud, Mazeaud & Donaldson, 1977, which are covered by the chapters written by N.W. Pankhurst and G. Van Der Kraak; G. McDonald and L. Milligan; C.B. Schreck, B.L. Olla and M.W. Davis; and T.G. Pottinger and A.D. Pickering, this volume). Most of what follows might be taken to suggest that all fish respond to stress with a fixed pattern of endocrine changes. However, this is manifestly not the case; a wide range of factors affect the endocrine response of a fish to stress. For example, external (environmental) factors such as prior exposure to pollutants (Barton, Weiner & Schreck, 1985; Hontela et al., 1992), or prophylactics (Pickering & Pottinger, 1985) affect the stress response, as does water quality (Pickering & Pottinger, 1987) and temperature (Perry & Reid, 1994). Besides the influences of such external factors, internal factors, such as sexual maturity, also modify the endocrine response to stress (Sumpter, Dye & Benfey, 1986; Sumpter et al., 1987; Pottinger, Balm & Pickering, 1995). Furthermore, in addition to these various factors which modify the stress response are endogenous endocrine rhythms, both daily and seasonal, in the concentrations of at least some of the stress hormones (Bry, 1982; Pickering & Pottinger, 1983). Hence, what follows provides no more than an insight into the endocrine responses that may well accompany a stress response; a very wide range of factors will influence which endocrine responses occur, and also the magnitude of these responses.

The adrenergic response to stress

It is very well documented that stress in fish is accompanied by rapid changes in the plasma concentrations of catecholamines, primarily adrenaline and noradrenaline. A wide variety of different stressors, including hypoxia (Butler *et al.*, 1978; Boutilier *et al.*, 1988; Perry & Reid, 1992), acidosis (Perry *et al.*, 1989; Brown & Whitehead, 1995), capture (Mazeaud *et al.*, 1977) and exhaustive exercise (Nankano & Tomlinson, 1967; Wood *et al.*, 1990) have been shown to lead to a rapid elevation in plasma concentrations of catecholamines. The catecholamines come primarily from the chromaffin cells scattered throughout the kidney and the walls of the posterior cardinal vein (Nilsson, 1984; Hathaway & Epple, 1989; Reid, Furimsky & Perry, 1994). Various physiological stimuli have been shown to trigger the mobilization of catecholamines from chromaffin cells, including the release of acetylcholine from pre-ganglionic fibres of the sympathetic nervous system (Perry *et al.*, 1991), the release of serotonin from neurones directly innervating the chromaffin cells (Fritsche *et al.*, 1993) and direct local effects of altered blood chemistry, such as disturbed acid–base balance or low arterial oxygen tension (Hathaway, Brinn & Epple, 1989; Perry & Reid, 1992).

Although catecholamines are released from stores within the chromaffin cells and posterior cardinal vein, acute stress does not lead to a decrease in the catecholamine content of these tissues, suggesting that biosynthesis is also stimulated by stress (Reid *et al.*, 1994). The ability to maintain catecholamine stores at 'normal' levels, despite responding to stress, ensures that sufficient amounts are available for an appropriate response, should a further stress occur. Nevertheless, there is some evidence available suggesting that repeated stress can lead to desensitization, although the mechanism(s) is unknown (Reid *et al.*, 1994). Further, long-term deterioration in the environment, such as that caused by depletion of oxygen, reduces the capacity to store catecholamines (Nilsson, 1990; Nilsson & Block, 1991), which obviously could affect the ability to mobilize catecholamines during a subsequent response to an acute stress.

The release of the catecholamines initiates a series of integrated responses, the aim of which is to ameliorate the disruptive effects of acute stress on normal physiology. In particular, the catecholamines serve to optimize cardiovascular and respiratory functions while also mobilizing energy stores for the increased metabolic requirements that are inevitably associated with stress (reviewed by Randall & Perry, 1992).

In different species of fish, the ratios of adrenaline: noradrenaline stored in the kidney and cardinal vein can vary; for example, the ratio appears to be around 1 : 1 in the brown trout, but 5 : 1 in the rainbow trout. Basal (unstressed) plasma concentrations of both catecholamines are in the low nanomolar range; they are usually reported to be between 0.5 and 10 nM (Gamperl, Vijayan & Boutilier, 1994; Brown & Whitehead, 1995, and references therein). They rise very rapidly after stress, as would be expected of a response involving neurally stimulated release of pre-stored hormone. If the stress is of very brief duration, plasma concentrations can fall equally rapidly (the half-life of catecholamines in the circulation is a few minutes) but, if the stress is more prolonged, plasma catecholamine concentrations can remain high for lengthy periods, such as for many days (Mazeaud *et al.*, 1977; Mazeaud & Mazeaud, 1981; Brown & Whitehead, 1995). Peak adrenaline and noradrenaline concentrations after stress appear to be about 100 nM (see, for example, Tang & Boutilier, 1988). However, the practical difficulties associated with collecting precisely timed blood samples immediately following initiation of an acute stress (for example, 1, 2, 3, 4, etc. minutes after the initial disturbance associated with a stress) mean that it is very difficult to define the exact time-course of the response and, therefore, to ascertain the maximum plasma concentrations attained [the same reasoning applies to changes in the plasma adrenocorticotrophin (ACTH) concentration in response to stress – see following section].

The 'simple' hypothalamo–pituitary–interrenal response to stress

Stress in fish, as in other animals, results in two types of endocrine response, the adrenergic response (which has already been discussed), resulting in increased plasma concentrations of adrenaline and noradrenaline, and the hypothalamo–pituitary–interrenal (HPI) response, culminating in an increased plasma cortisol concentration.

The HPI axis is a cascade of hormones, with cortisol as the physiologically important hormone responsible for the effects of stress. The first hormone in this cascade is corticotrophin-releasing hormone, or CRH. Neurones secreting CRH hold a position of particular importance in mediating the response to stress, because not only do they influence cortisol concentrations through the classical HPI axis, but they also moderate the release of catecholamines from the sympathetic nervous system, and affect the expression of various behavioural patterns through central pathways. CRH, a peptide, has not been purified and characterized from any species of fish. However, its sequence can be deduced if the nucleotide sequence of the gene is known. Unfortunately, the nucleotide sequence for fish CRH messenger ribonucleic acid (mRNA) has been established only for one species, the white sucker *Catostomus commersoni*. This species is considered tetraploid, and therefore two (or more) distinct genes coding for CRH might be expected (the same is true of salmonids, and of a number of other families of fish). In fact, two distinct genes for CRH, which are both transcribed into distinct mRNAs encoding CRH polypeptide precursors, have been identified and characterized (Okawara *et al.*, 1988; Morley *et al.*, 1991). The two sucker CRH precursors are organized in a similar way and, in turn, are similar to the rat CRH precursor. Potentially, they can be processed to give rise to two mature CRH peptides, each with a length of 41 amino acid residues, which differ by just one residue, an Ala to Val substitution. Furthermore, sucker CRH 1 and 2 show, repetively, only two and three amino acid differences from the rat sequence (Morley *et al.*, 1991). Thus, there has been extreme conservation in these CRH genes for at least 300 million to 400 million years, since the separation of the lines leading to present-day teleosts and mammals.

The existence of CRH neurones in fish has been demonstrated using immunocytochemistry (but not yet by *in situ* histochemistry using a probe to CRH mRNA). Use of antiserum to mammalian CRH (which probably binds fish CRH – no antiserum to fish CRH is available) has allowed the identification of a CRH-like system in a number of species of fish (Yulis *et al.*, 1986; Olivereau & Olivereau, 1988). Parvocellular and magnocellular perikarya were labelled in the preoptic nucleus, and their labelled fibres were traced laterally, as they formed two symmetrical tracts running through the basal hypothalamus. These tracts ended in the rostral pars distalis, close to corticotrophs (ACTH-secreting cells). Other fibres ended in the neurointermediate lobe, close to melanotrophs. In some species, small perikarya also stained in the nucleus lateralis tuberis (Olivereau & Olivereau, 1988).

The main role of CRH, at least in the HPI axis, appears to be the control of ACTH release from the corticotrophs of the anterior pituitary gland. However, despite this central role of CRH in the HPI axis, very little research has directly addressed the control of secretion of ACTH in fishes by CRH. The reasons for this are numerous, but of major significance is the fact that, unlike in mammals and other higher vertebrates, CRH reaches the cells of the pituitary gland via direct neural contact, instead of via a portal blood system. Hence, administering CRH to a fish in such a way that it reaches the corticotrophs and can gain access to CRH receptors is likely to be very difficult, if not impossible. However, mammalian CRH does appear able to stimulate the release of ACTH from isolated pituitary cells and cultured pituitary glands (Fryer & Lederis, 1986; Baker, Bird & Buckingham, 1985), suggesting that it is indeed a physiological regulator of ACTH secretion.

ACTH, like CRH, is synthesized as part of a larger precursor. The precursor, termed proopiomelanocortin (POMC), is the common precursor of not only ACTH, but also β-endorphin, the melanotrophins (MSHs) and other peptides. POMC is synthesized in two cell types of the pituitary gland, the corticotrophs of the pars distalis and the melanotrophs of the neurointermediate lobe.

The complementary deoxyribonucleic acids (cDNAs) for two quite different POMCs in the rainbow trout have been characterized; this is likely a reflection of the tetraploid nature of salmonids (see prior discussion on two CRHs). These cDNAs exhibit only limited sequence homology (44%) and their deduced amino acid sequences also show only weak similarity (43%), despite the high degree of conservation of some short stretches within the precursors (Salbert *et al.*, 1992). A single POMC cDNA from chum salmon (*Oncorhynchus keta)* has also been characterized (Soma *et al.*, 1984; Kitahara *et al.*, 1988), as has a POMC-like cDNA from the lamprey *Petromyzon marinus*, one of the most phylogenetically ancient agnathans with a lineage which can be traced back approximately 550 million years (Heinig *et al.*, 1995).

All of these POMCs have the same structural organization. In the corticotrophs, POMC is thought to be enzymatically cleaved very specifically to yield an N-terminal peptide (sometimes called NPP), ACTH, and β-lipotrophin. It is likely, although not proven, that all of these peptides are secreted concomitantly from the corticotrophs (Rodrigues & Sumpter, 1983). The biological activities of many of these peptides remain uncertain; for example, neither the N-terminal peptide nor β-lipotrophin have any well-defined roles. However, the role of ACTH in controlling synthesis and release of cortisol from the interrenal tissue is very well documented (Ilan & Yaron, 1980).

The two distinct POMCs in trout code for two sequentially different ACTHs; both consist of about 40 amino acid residues, differing primarily at the C-terminal end (this end of ACTH is thought to be less important for biological activity than the N-terminal end of the hormone; Rance & Baker, 1981). Because the actual ACTH peptides (rather than their cDNAs) have not been purified and characterized, it is not possible to be certain as to exactly how the POMC precursors are processed, or what the exact sequences of the ACTHs are that are secreted. Furthermore, although it seems likely that two structurally distinct ACTHs are secreted (at least by tetraploid fish), this has yet to be demonstrated.

ACTH is secreted very rapidly in response to stress; within 3 min of an acute stress the plasma ACTH concentration had risen significantly in trout, and the peak concentration occurred after only 5 min (Sumpter *et al.*, 1986). A variety of studies of trout have shown that the first measurable response of the HPI axis to stress is an increase in the plasma ACTH concentration (Sumpter *et al.*, 1986; Pickering *et al.*, 1986; Pickering, Pottinger & Sumpter, 1987*a*). In contrast, in tilapia *(Oreochromis mossambicus)*, stress may not always lead to an increase in the plasma ACTH concentration (Balm *et al.*, 1994). However, plasma ACTH concentrations are difficult to measure (hence the limited data available) because of the high degree of specificity and sensitivity required; therefore, the existing data, particularly those on 'basal' concentrations (which are reported to be about 20 pg/ml), should be interpreted with caution. It is my personal opinion that, when better ACTH assays are available (such as those based on two antibodies – the so-called immunoradiometric assays or IRMAs, which provide both increased specificity and sensitivity), the 'basal' ACTH concentration will be lower than existing reports suggest. Development and use of such assays may well show changes in plasma concentrations that are obscured by existing assays, due to their lack of appropriate sensitivity. Based on existing data, plasma concentrations of ACTH in unstressed fish appear to be around 20–30 pg/ml, rising to many hundred picogram per millilitre in stressed fish. If fish are chronically stressed (for many days), plasma ACTH concentrations may fall back to 'basal' levels, accompanied by a concomitant fall in plasma cortisol concentrations (Balm & Pottinger, 1995), suggesting that habituation to stress (see next paragraph) is regulated at or above the level of the pituitary gland, via plasma ACTH concentrations.

ACTH acts on the interrenal tissue to stimulate the release of corticosteroids, particularly cortisol, which is the predominant corticosteroid in most of the (teleost) fishes studied so far (Henderson & Garland, 1980). Cortisol is not stored to any significant extent in interrenal tissue, but instead is synthesized as required. Basal cortisol concentrations in unstressed fish are generally reported to be between <1 and 10 ng/ml (Sumpter *et al.*, 1986; Pottinger & Moran, 1993) and rise fairly rapidly in response to stress. Generally, the first detectable rise occurs around 5–10 min after initiation of the stress (factors such as water temperature affect the rate of response), and concentrations continue to rise, as long as the stress is sustained, to reach between 100 and 200 ng cortisol/ml after about 1 h (Sumpter, Pickering &

Pottinger, 1985; Sumpter *et al.*, 1986; Pickering & Pottinger, 1989; Pottinger & Moran, 1993), although substantially higher concentrations have been reported (see, for example, Brown & Whitehead, 1995). In the case of acute stresses, such as handling, brief disturbance or short-term confinement, elevation of plasma cortisol concentrations may last for only a matter of a few hours (Pickering & Pottinger, 1989). In contrast, when fish are subjected to long-term (chronic) or continuous stresses, such as prolonged confinement, crowding or subordination, plasma cortisol concentrations can be elevated for many days, or even weeks (Pickering & Pottinger, 1989). In some, but apparently not all, instances of prolonged stress, the plasma cortisol concentration may eventually return to basal values (Pickering & Stewart, 1984) despite the continued stress; this demonstrates acclimation (habituation) to the stressful conditions.

Although, as stated earlier, cortisol is considered the major corticosteroid in most fish (certainly most teleosts), other corticosteroids are present in appreciable amounts. For example, it has recently been shown that plasma cortisone concentrations rise in response to stress (Pottinger & Moran, 1993). The rate of increase of the plasma cortisone concentration during stress was actually more rapid than that of cortisol, with maximum values (100–200 ng/ml, and hence similar to cortisol concentrations) being reached within 10–20 min of the beginning of the stress. The physiological significance of this cortisone response is unknown; although it may simply reflect a rapid conversion of cortisol to cortisone (Pottinger & Moran, 1993), this does not mean that it is without effect.

As far as the physiological consequences of activation of the HPI axis are concerned, a considerable amount of evidence suggests that these (such as impaired reproduction, reduced growth rate, increased susceptibility to disease) are a direct consequence of the elevated cortisol concentrations that occur in response to stress. However, other mechanisms may be partly, or even completely, responsible (for example, see chapter by N.W. Pankhurst and G. Van Der Kraak, this volume). Thus, both CRH and ACTH appear to have only one function in the HPI axis – to control the plasma concentration of cortisol, and only cortisol is responsible for the effects of stress. The effects, which are both wide ranging and very important (they can be life-threatening), are dealt with in depth by other chapters in this book.

The 'complex' HPI axis

The HPI axis, as described above, seems fairly simple: central mechanisms lead to secretion of CRH, which stimulates ACTH secretion into the blood, which in turn stimulates cortisol synthesis and secretion from the interrenal tissue. However, the HPI axis is not as simple as that; a variety of other hormones interact with CRH and ACTH at all levels in ultimately controlling the blood cortisol concentration. Already the number of factors (primarily hormones) reported to modify the actions of CRH and ACTH is quite large, and is becoming steadily larger; simply listing them all here would serve little purpose. Instead, I intend to provide a few examples, which illustrate the complexity observed at all levels of the HPI axis (see also Fig. 1).

Factors interacting with CRH in the control of ACTH secretion

As stated earlier (p. 98), CRH-secreting neurones play a pivotal role in the control of ACTH release. However, other factors appear to interact with CRH to varying degrees. For example, it is very clear that in mammals CRH acts synergistically with vasopressin in stimulating ACTH release, and that vasopressin is co-expressed in the CRH neurones. A similar interaction between these two neuropeptides in controlling ACTH secretion probably exists in fish also. Co-localization of CRH and vasopressin has been demonstrated to occur in the brain and pituitary gland of fish (Yulis & Lederis, 1987; Olivereau & Olivereau, 1990) and vasopressin can stimulate ACTH secretion (Fryer & Lederis, 1986). Perhaps, as demonstrated to occur in mammals, CRH and vasopressin act synergistically in stimulating the release of ACTH from the fish pituitary gland.

A number of other neuropeptides have been shown to interact with CRH in controlling the secretion of ACTH, although exactly how this interaction occurs is unclear. For example, melanin-concentrating hormone (MCH) depresses ACTH (and hence corticosteroid) release during stress, acting via central pathways that presumably involve reduced CRH release (Green, Baker & Kawauchi, 1992). Similarly, urotensin I and sauvagine, two peptides that are structurally closely related to CRH, can stimulate the release of ACTH from the corticotrophs (and possibly also the melanotrophs) of the goldfish (Fryer, Lederis & Rivier, 1983; Tran et al., 1990). It is not known whether sauvagine, a peptide isolated initially from the skin of amphibians, is present in fish, but urotensin I is found in many species of fish (Lederis et al., 1982; Waugh et al., 1995), where it is located in many areas of the central nervous system (CNS), including the brain and pituitary gland (McMaster & Lederis, 1989) and, hence, obviously could interact with CRH neurones and/or interact with CRH at the corticotrophs.

HIGHER BRAIN CENTRES

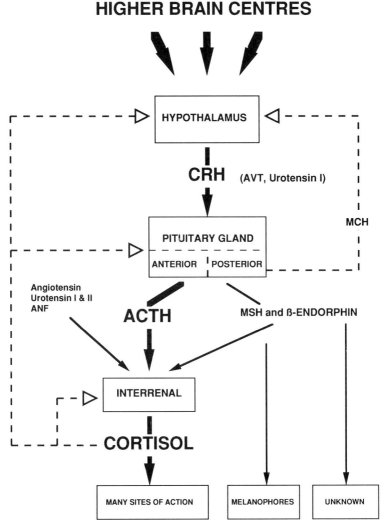

Fig. 1. The hypothalamo–pituitary–interrenal axis of fish. This is definitely a simplified version, but nevertheless still serves to illustrate how complex the axis is. The hormones (defined in the text) thought presently to play major roles are shown in bold. Solid arrows (and arrowheads) represent positive (stimulatory) actions, whereas broken arrows (and open arrowheads) represent negative (inhibitory) actions; the latter are often referred to as negative feedback.

Factors interacting with ACTH in the control of cortisol secretion

It seems clear that ACTH plays the central role in controlling secretion of corticosteroids from the interrenal tissue. However, many other peptides appear to be able to influence interrenal secretion of cortisol in fish. One group of related hormones that has received a reasonable amount of attention is that derived from the precursor protein POMC. As discussed earlier (p. 99), this precursor is synthesized in both the corticotrophs of the anterior pituitary gland and the melanotrophs of the posterior pituitary gland (the neurointermediate lobe). POMC appears to be processed further in the melanotrophs than it is in the corticotrophs (Rodrigues & Sumpter, 1983), leading to the secretion of somewhat different sets of peptides from the two cell types. As far as one can judge presently, ACTH is the major biologically active product from the processing of POMC in the corticotrophs, whereas α-MSH and β-endorphin are major products from the melanotrophs (although their actions are less well defined).

Some, but probably not all, types of stress cause the secretion of peptides from the melanotrophs as well as from the corticotrophs. Thus, Sumpter et al. (1985) showed that plasma concentrations of α-MSH and β-endorphin, as well as cortisol (as expected), were raised in brown trout subjected to a fairly severe stress. Likewise, a rise in plasma α-MSH concentrations accompanied that of ACTH and cortisol when rainbow trout were restrained out of water (Sumpter et al., 1986). Acidified water (presumed to be a stress) activates the melanotrophs and leads to an increase in plasma MSH concentrations in tilapia (Lamers et al., 1991). However, other types and/or degrees of stress have led to either no change (Pickering et al., 1986) or even a reduction in plasma MSH concentrations (Balm & Pottinger, 1995). The functional significance of this release of peptides derived from processing of POMC in melanotrophs is unclear. However, it has been suggested that the secretion of MSH and β-endorphin from the melanotrophs of tilapia (O. mossambicus) during some forms of stress may act in a co-ordinated manner and interact with ACTH in the control of cortisol secretion (Balm, Hovens & Wendelaar-Bonga, 1995). Certainly α-MSH and β-endorphin can act in an additive (or possibly even synergistic) manner to stimulate release of cortisol from interrenal tissue in culture – and this effect occurs at physiologically meaningful concentrations of the two peptides. Overall, the evidence to date (Balm et al., 1995) suggests that products from the melanotrophs can be potent regulators of corticosteroidogenesis in fish, although the mechanism does not appear to operate in response to all types and/or degrees of stress.

Besides the potential role of other peptides (in additon to ACTH) derived from POMC in regulating plasma cortisol concentrations, other quite unconnected peptide hormones have also been reported to influence the secretion of cortisol. For example, angiotensin and both urotensins (I and II) were able to stimulate cortisol secretion *in vitro* and were able, also, to enhance the stimulating action of ACTH. In most cases, the responses to a combination of ACTH and another peptide hormone were additive, but occasionally they were synergistic (Arnold-Reed & Balment, 1994). Atrial natriuretic factor (ANF) has similar effects on cortisol secretion (Arnold-Reed & Balment, 1991).

Feedback mechanisms

As currently presented, the HPI axis, whether considered in a 'simple' or 'complex' way, consists of a cascade of hormones leading to the regulation of cortisol secretion from the interrenal tissue. In fact, this is not so; instead, feedback mechanisms operate to keep the axis in check. All the feedback loops reported to occur in the HPI axis of fish are negative ones, meaning that they suppress one or more steps in the cascade of hormones. These feedback loops are often described as short, in which a hormone may directly regulate its own secretion, or long, in which the final product of the cascade acts back at steps earlier in the cascade. Both types of feedback loops have been described as being part of the HPI axis, although relatively little information is available about either (see Fig. 1 for some examples).

Most of the information that is available refers to the ability of cortisol to exert some degree of negative feedback on its own rate of secretion. Thus, cortisol can inhibit the ability of CRH (and other putative secretogogues such as urotensin I and sauvagine) to stimulate ACTH release from cultured pituitary cells (Fryer, Lederis & Rivier, 1984), and dexamethasone (a synthetic corticosteroid agonist) can inhibit the elevations of plasma ACTH and cortisol concentrations that normally occur in response to stress (Pickering *et al.*, 1987a). Furthermore, sexually mature salmonids, which have chronically elevated plasma cortisol concentrations, show a blunted ACTH response when acutely stressed (Sumpter *et al.*, 1986). Taken together, these results suggest that cortisol acts back at both the hypothalamus and pituitary gland to inhibit CRH and ACTH release, and thereby dampen the response to stress.

Much less information is available on whether short-loop feedback mechanisms operate as a normal part of the HPI axis. However, some evidence for their existence was provided by Bradford, Fitzpatrick & Schreck (1992), who showed that, in the presence of cortisol, interrenal tissue was less sensitive to stimulation by ACTH than when it was absent. Thus, cortisol appears able to suppress its own secretion.

Despite all the complexities, only some of which are discussed here, it is important to emphasize the absolutely central role of CRH, ACTH and cortisol in any response of the HPI axis to stress. Although many other hormones may play modulatory roles, it still appears true to say that CRH plays the most important role in regulating ACTH release from the corticotrophs of the anterior pituitary gland and that, in turn, ACTH plays by far the major role in regulating cortisol secretion from the interrenal tissue (Fig. 1).

Effects of stress on other hormones

Growth hormone

Stress leads to suppression of growth (Pickering, 1990), although the exact mechanism(s) responsible remains unclear (see also chapter by N.W. Pankhurst and G. Van Der Kraak). There are relatively few reports on plasma growth hormone (GH) concentrations in reponse to stress, and those that there are do not provide an unequivocal picture. Thus, an increase (Cook & Peter, 1984; Takahashi et al., 1991), a decrease (Pickering et al., 1991; Farbridge & Leatherland, 1992), or no change (Wagner & McKeown, 1986) in the plasma GH concentration in response to stress have all been reported. Such apparently contradictory results could be due to differences in the type and degree of stress, other differences in experimental design, species differences, or technical problems, or any combination of these factors!

Prolactin

The effects of stress on plasma prolactin concentrations in fish have been studied relatively little, and those studies which have been conducted report somewhat contradictory results. Both increased (Avella, Schreck & Prunet, 1991; Betoulle et al., 1995) and decreased (Pottinger, Prunet & Pickering, 1992) prolactin concentrations in response to stress have been reported. As already stated (when discussing effects on the plasma GH concentration), there are many possible reasons to account for these variable responses.

Gonadotrophins

The continued difficulty in measuring plasma concentrations of the two gonadotrophins [GTHI and GTHII, which are analagous to follicle-stimulating hormone (FSH) and luteinizing hormone (LH) of higher vertebrates, respectively] has prevented studies on the effects of stress on each individual gonadotrophin. However, assays of somewhat uncertain specificity have been used to investigate whether stress affects gonadotrophin concentrations. The few reports available do not provide a consistent picture; for example, whereas Gillet, Billard & Breton (1981) reported that chronic stress resulted in a decrease in both plasma and pituitary gonadotrophin levels in goldfish, Pickering *et al.* (1987*b*) found that acute stress led to an increase in the plasma gonadotrophin concentration. Chronic elevation of the plasma cortisol concentration (achieved by use of a cortisol-releasing implant) led to a reduced content of gonadotrophin in the pituitary gland and lowered plasma concentration (Carragher *et al.*, 1989).

Somatolactin

Relatively little is known about this pituitary hormone presently, because it was only discovered recently. However, plasma concentrations are known to rise very rapidly in response to stress (Rand-Weaver, Pottinger & Sumpter, 1993; Kakizawa *et al.*, 1995). The timing of this response – peak concentrations are reached within minutes – strongly suggests direct stimulation of the somatolactin-secreting cells via the hypothalamus. This suggestion is supported by the observation that neurones staining positively for CRH project from the hypothalamus into the posterior pituitary gland, and end on the cells which secrete somatolactin (Olivereau & Olivereau, 1988, 1990).

The examples above, which represent only some of many that could have been chosen, demonstrate the very wide-ranging endocrine effects of stress (Sumpter *et al.*, 1994). Although all of the examples I have chosen are pituitary hormones, stress undoubtedly also affects the secretion of hormones from other endocrine tissues. The effects of stress on these hormones will, in turn, affect a wide range of physiological processes. Because some of the hormones mentioned in the previous sections are themselves part of a hormone cascade, then it is likely that the entire cascade is affected. For example, GH does not act directly; instead most, if not all, of the actions of GH are mediated indirectly via the control GH exerts on the secretion of insulin-like growth factor I (IGF-I). Thus, although not yet demonstrated, it is likely that stress, by affecting plasma GH concentrations, also affects plasma IFG-I concentrations, and it is these which affect the growth rate (and the other processes influenced by IGF-I).

A second example of the wide-ranging effects of stress on the endocrine system would be the consequences of altered gonadotrophin secretion in reponse to stress. As already stated, chronically elevated cortisol concentrations lead to reduced pituitary and plasma gonadotrophin levels (Pickering *et al.*, 1987*b*; Carragher *et al.*, 1989), which, in turn, lead to reduced androgen concentrations in male fish (Pickering *et al.*, 1987*b*; Melotti *et al.*, 1992), and reduced oestrogen and 17α,20β-dihydroxyprogesterone concentrations in female fish (Carragher *et al.*, 1989; Foo & Lam, 1993; Pankhurst & Dedual, 1994). All of these changes collectively result in smaller gonads, the production of fewer gametes, and a reduced quality of those gametes which are produced (Campbell, Pottinger & Sumpter, 1992, 1994).

Interaction of adrenergic and HPI responses to stress

As portrayed in this chapter so far, activation of the adrenergic axis and the HPI axis appear as two quite distinct, parallel responses to stress. However, recent evidence suggests that there is cross-talk (interaction) between these two axes, with each one modulating the other. Relatively little research has been focused on this possible cross-talk (because researchers tend to focus on either the HPI axis or the adrenergic response to stress), but the little that is known does suggest that a significant degree of interaction occurs.

The ability of adrenaline, administered *in vivo*, to cause an increase in the plasma cortisol concentration (White & Fletcher, 1985; Gamperl *et al.*, 1994) suggests that the elevated adrenaline concentrations associated with a stress response may stimulate the HPI axis. In fish, the mechanism of this stimulation is unknown but, in mammals, a similar interaction between the two endocrine axes stimulated by stress is mediated by adrenaline acting via adreno-receptors in both the hypothalamus and pituitary gland to increase CRH secretion and ACTH release (Plotsky, Cunningham & Widmaier, 1989).

Not only does it appear that the adrenergic response can enhance the activation of the HPI axis, but it also appears that the reverse occurs, with activation of the HPI axis during stress enhancing the adrenergic response. Thus, for example, chronic elevation of the cortisol concentration seems to increase the sensitivity of the β-adrenergic signal transduction system (by causing an increase in the number of β-adrenoreceptors) of at least two cell types which play major roles in the physiological responses to stress – the red blood cells (RBCs) and the hepatocytes (Perry & Reid, 1993). This increase in receptor number on RBCs enhances a series of physiological and biochemical mechanisms which alleviate the disruptive effects of stress on blood oxygen transport. Similarly, enhancement of cell surface β-adrenoreceptors by cortisol increases the physiological responsiveness (glycogenolysis) of hepatocytes to catecholamines (reviewed in Perry & Reid, 1993). Furthermore, the increase in plasma cortisol concentration that occurs in reponse to stress also appears to accelerate the rate of catecholamine biosynthesis (Jönsson, Wahlquist & Hansson, 1983; Nilsson, 1984).

Thus, although the existing knowledge is rather fragmentary, it does appear that the two endocrine axes activated by stress interact in various ways to enhance each other, and thereby alleviate the adverse effects of stress as rapidly as possible.

Conclusions

The endocrine changes associated with stress are wide ranging; it is doubtful whether any endocrine pathway is unaffected during a stress response. Although this review possibly portrays a much more complex picture than the reader might have expected, in fact it probably underestimates, very significantly, the complexities of the endocrine changes associated with a stress response. There are still major gaps in our knowledge, some of which include:

1. What endocrine changes are occurring in the CNS, especially the brain (where stress is perceived) which lead to activation of the HPI and catecholamine pathways?

This is obviously an extremely complex area. Presently, and quite understandably, the only information available relates to gross changes in neurotransmitter content of the brain as a whole. Unsurprisingly, it was reported that the amount in the brain of many different neurotransmitters in the teleost *Sciaenops ocellatus* was affected by stress (Ortiz & Lutz, 1995). There was also some indication that different types of stress might involve different responses.

2. What is occurring at the level of the receptor?

All hormones act through receptors. Very little is known about the relevant receptors in fish. Thus, for example, whereas the CRH, ACTH, glucocorticoid and adrenergic receptors (multiple types of some) have been characterized in some higher vertebrates (usually down to the molecular level), this has not been achieved for any species of fish.

Therefore, we presently know absolutely nothing about the receptors for CRH and ACTH, and only very little about the receptors for the catecholamines and glucocorticoids (see, for example, Maule & Schreck, 1991; Pottinger, Knudson & Wilson, 1994; Shrimpton & Randall, 1994).

Much more research on these receptors is required, particularly on their molecular characterization. Not only would such structural knowledge be very informative in itself, but the 'tools' obtained (such as mRNA sequences coding for the receptors) would make possible a whole range of studies of receptor dynamics which cannot be done at the present time. It should be remembered that receptors are not static structures, but instead most of their parameters, such as their affinity, location within the cell, turnover and abundance, are all very fluid, and would be expected to change during a stress response; in fact, changes in some of these characteristics of glucocorticoid receptors during a stress response have already been studied (Maule & Schreck, 1991; Pottinger *et al.*, 1994; Shrimpton & Randall, 1994).

3. How much of the endocrine system is affected by stress?

My personal opinion in answer to this question would be 'all of it'. Besides the obvious activation of the adrenergic and HPI axes, at which most research on the endocrinology of stress has been directed, limited studies have shown that the plasma concentrations of many other hormones are perturbed; for example, it seems likely that the secretion of all pituitary hormones is affected by stress (only in the case of the neurohypophysial hormones is there no information). Nothing appears to be known in regard to the (possible) effects of stress on such major groups of hormones as the gastrointestinal (such as gastrin, secretin, vasoactive intestinal polypeptide) and pancreatic (insulin, glucagon, somatostatin) hormones and the hormones controlling calcium homeostatis (stanniocalcin).

It will take an awful lot of research to answer these, and many other outstanding, questions. Such research will, inevitably, generate a whole new set of questions: endocrinologists interested in the responses of fish to stress will not be short of issues to address.

References

Arnold-Reed, D.E. & Balment, R.J. (1991). Atrial natriuretic factor stimulates *in vivo* and *in vitro* secretion of cortisol in teleosts. *Journal of Endocrinology*, **128**, R17–R20.

Arnold-Reed, D.E. & Balment, R.J. (1994). Peptide hormones influence *in vitro* interrenal secretion of cortisal in the trout, *Oncorhynchus mykiss. General and Comparative Endocrinology*, **96,** 85–91.

Avella, M., Schreck, C.B. & Prunet, P. (1991). Plasma prolactin and cortisol concentrations of stressed coho salmon *(Oncorhynchus kisutch)* in freshwater or seawater. *General and Comparative Endocrinology*, **81**, 21–7.

Baker, B.I., Bird, D.J. & Buckingham, J.C. (1985). Salmonid melanin-concentrating hormone inhibits corticotrophin release. *Journal of Endocrinolopgy*, **106**, R5–R8.

Balm, P.H.M. & Pottinger, T.G. (1995). Corticotrope and melanotrope POMC-derived peptides in relation to interrenal function during stress in rainbow trout *(Oncorhynchus mykiss). General and Comparative Endocrinology*, **98**, 279–88.

Balm, P.H.M., Pepels, P., Helfrich, S., Hovens, M.L.M. & Wendelaar-Bonga, S.E. (1994). Adrenocorticotropic hormone (ACTH) in relation to interrenal function during stress in tilapia *(Oreochromis mossambicus). General and Comparative Endocrinology*, **96**, 347–60.

Balm, P.H.M., Hovens, M.L.M. & Wendelaar Bonga, S.E. (1995). Endorphin and MSH in concert form the corticotropic principle released by Tilapia *(Oreochromis mossambicus; Teleostei)* melanotropes. *Peptides*, **16**, 463–9.

Barton, B.A., Weiner, G.S. & Schreck, C.B. (1985). Effect of prior acid exposure on physiological responses of juvenile rainbow trout *(Salmo gairdneri)* to acute handling stress. *Canadian Journal of Fisheries and Aquatic Sciences*, **42**, 710–17.

Betoulle, S., Troutaud, D., Khan, N. & Deschaux, P. (1995). Antibody-response, cortisol and prolactin levels in rainbow trout. *Comptes Rendus de L'academie des Sciences. Serie III – Sciences de la Vie.* **318**, 677–81.

Boutilier, R.G., Dobson, G.P., Hoegar, U. & Randall, D.J. (1988). Acute exposure to graded levels of hypoxia in rainbow trout *(Salmo gairdneri)*: metabolic and respiratory adaptations. *Respiratory Physiology*, **71**, 69–82.

Bradford, C.S., Fitzpatrick, M.S. & Schreck, C.B. (1992). Evidence for ultra-short-loop feedback in ACTH-induced interrenal steroidogenesis in coho salmon: acute self-suppression of cortisol secretion *in vitro. General and Comparative Endocrinology*, **87**, 292–9.

Brown, J.A. & Whitehead, C. (1995). Catecholamine release and interrenal response of brown trout, *Salmo trutta*, exposed to aluminium in acidic water. *Journal of Fish Biology*, **46**, 524–35.

Bry, C. (1982). Daily variations in plasma cortisol levels of individual female rainbow trout, *Salmo gairdneri*: evidence for a post-feeding peak in well-adapted fish. *General and Comparative Endocrinology*, **48**, 462–8.

Butler, P.J., Taylor, E.W., Capra, M.F. & Davison, W. (1978). The effect of hypoxia on the levels of circulating catecholamines in the dogfish *(Scyliorhinus canicula)*. *Journal of Comparative Physiology*, **127**, 325–30.

Campbell, P.M., Pottinger, T.G. & Sumpter, J.P. (1992). Stress reduces the quality of gametes produced by rainbow trout. *Biology of Reproduction*, **47**, 1140–50.

Campbell, P.M., Pottinger, T.G. & Sumpter, J.P. (1994). Preliminary evidence that chronic confinement stress reduces the quality of gametes produced by brown and rainbow trout. *Aquaculture*, **120**, 151–69.

Carragher, J.F., Sumpter, J.P., Pottinger, T.G. & Pickering, A.D. (1989). The deleterious effects of cortisol implantation on reproductive function in two species of trout, *Salmo trutta* L. and *Salmo gairdneri*. *General and Comparative Endocrinology*, **76**, 310–21.

Cook, A.F. & Peter, R.E. (1984). The effects of somatostatin on serum growth hormone levels in the goldfish, *Carassius auratus*. *General and Comparative Endocrinology*, **54**, 109–13.

Farbridge, K.J. & Leatherland, J.F. (1992). Plasma growth hormone levels in fed and fasted rainbow trout *(Oncorhynchus mykiss)* are decreased following handling stress. *Fish Physiology and Biochemistry*, **10**, 67–73.

Foo, J.T.W. & Lam, T.J. (1993). Retardation of ovarian growth and depression of serum steroid levels in tilapia, *Oreochromis mossambicus*, by cortisol implantation. *Aquaculture*, **115**, 133–43.

Fritsche, R., Reid, S.G., Thomas, S. & Perry, S.F. (1993). Serotonin-mediated release of catecholamines in the rainbow trout, *(Oncorhynchus mykiss)*. *Journal of Experimental Biology*, **178**, 191–204.

Fryer, J.N. & Lederis, K. (1986). Control of corticotropin secretion in teleost fishes. *American Zoologist*, **26**, 1017–26.

Fryer, J.N., Lederis, K. & Rivier, J. (1983). Urotensin I, a CRF-like peptide, stimulates ACTH release from the teleost pituitary. *Endocrinology*, **113**, 2308–10.

Fryer, J.N., Lederis, K. & Rivier, J. (1984). Cortisol inhibits the ACTH-releasing activity of urotensin I, CRF and sauvagine observed with superfused goldfish pituitary cells. *Peptides*, **5**, 925–30.

Gamperl, A.K., Vijayan, M.M. & Boutilier, R.G. (1994). Epinephrine, norepinephrine, and cortisol concentrations in cannulated seawater-acclimated rainbow trout *(Oncorhynchus mykiss)* following black-box confinement and epinephrine injection. *Journal of Fish Biology*, **45**, 313–24.

Gillet, C., Billard, R. & Breton, B. (1981). La reproduction du poisson rouge *Carassius auratus* élev́ à 30 °C : effet de la photopériode, de l'alimentation et de l'oxygénation. *Cahiers Laboratoire Montereau*, **11**, 49–56.

Green, J.A., Baker, B.I. & Kawauchi, H. (1992). The effect of rearing rainbow trout on black or white backgrounds on their secretion of melanin-concentrating hormone and their sensitivity to stress. *Journal of Endocrinology*, **128**, 267–74.

Hathaway, C.B. & Epple, A. (1989). The sources of plasma catecholamines in the American eel, *Anguilla rostrata*. *General and Comparative Endocrinology*, **74**, 418–30.

Hathaway, C.B., Brinn, J.E. & Epple, A. (1989). Catecholamine release in the eel does not require the presence of brain or anterior spinal cord. *Journal of Experimental Zoology*, **249**, 338–42.

Heinig, J.A., Keeley, F.W., Robson, P., Sower, S.A. & Youson, J.H. (1995). The appearance of proopiomelanocortin early in vertebrate evolution: cloning and sequencing of POMC from a lamprey pituitary cDNA library. *General and Comparative Endocrinology*, **99**, 137–44.

Henderson, I.W. & Garland, H.O. (1980). The interrenal gland in pisces. Part 2. Physiology. In *General, Comparative and Clinical Endocrinology of the Adrenal Cortex,* Volume 3. pp. 474–523. Academic Press, London.

Hontela, A., Rasmussen, J.B., Audet, C. & Chevalier, G. (1992). Impaired cortisol stress response in fish from environments polluted by PAHs, PCBs, and mercury. *Archives of Environmental Contamination and Toxicology*, **22**, 278–83.

Ilan, Z. & Yaron, Z. (1980). Stimulation of cortisol secretion *in vitro* from the interrenal tissue of the cichlid fish, *Sarotherodon aureus*, by adrenocorticotrophin or cyclic AMP. *Journal of Endocrinology*, **86**, 269–77.

Jönsson, A.C., Wahlquist, I. & Hansson, T. (1983). Effects of hypophysectomy and cortisol on the catecholamine biosynthesis and catecholamine content in chromaffin tissue from rainbow trout, *Salmo gairdneri*. *General and Comparative Endocrinology*, **51**, 278–85.

Kakizawa, S., Kaneko, T., Hasegawa, S. & Hirano, T. (1995). Effects of feeding, fasting, background adaptation, acute stress, and exhaustive exercise on the plasma somatolactin concentrations in rainbow trout. *General and Comparative Endocrinology*, **98**, 137–46.

Kitahara, N., Nishizawa, T., Iida, K., Okazaki, H., Andoh, T. & Soma, G.-I. (1988). Absence of γ-melanocyte-stimulating hormone sequence in proopiomelanocortin mRNA of chum salmon, *Oncorhynchus keta*. *Comparative Biochemistry and Physiology*, **91B**, 365–70.

Lamers, A.E., Balm, P.H.M., Haenan, H.E.M.G. & Wendelaar-Bonga, S.E. (1991). Regulation of differential release of α-melanocyte-stimulating hormone forms from the pituitary of teleost fish, *Oreochromis mossambicus*. *Journal of Endocrinology*, **129**, 179–87.

Lederis, K., Letter, A., McMaster, O., Moore, G. & Schlesinger, D. (1982). Complete amino acid sequence of urotensin I, a hypotensive and corticotropin-releasing neuropeptide, from *Catostomus*. *Science*, **218**, 162–4.

Maule, A.G. & Schreck, C.B. (1991). Stress and cortisol treatment changed affinity and number of glucocorticoid receptors in leukocytes and gill of coho salmon. *General and Comparative Endocrinology*, **84**, 83–93.

Mazeaud, M.M. & Mazeaud, F. (1981). The role of catecholamines in the stress response in fish. In *Stress and Fish*. Pickering, A.D. (ed.) pp. 49–75. Academic Press, London.

Mazeaud, M.M., Mazeaud, F. & Donaldson, E.M. (1977). Primary and secondary effects of stress in fish: some new data with a general review. *Transactions of the American Fisheries Society*, **106**, 201–12.

McMaster, D. & Lederis, K. (1989). Urotensin I and CRF-like peptides in *Catostomus commersoni* brain and pituitary: HPLC and RIA characterization. *Peptides*, **9**, 1043–8.

Melotti, P., Roncarati, A., Garella, E., Carnevali, O., Mosconi, G. & Polzonetti-Magni, A. (1992). Effects of handling and captive stress on plasma glucose, cortisol and androgen levels in brown trout, *Salmo trutta morpha fario*. *Journal of Applied Ichthyology*, **8**, 234–9.

Morley, S.D., Schnrock, C., Richter, D., Okawara, Y. & Lederis, K. (1991). Corticotropin-releasing factor (CRF) gene family in the brain of the teleost fish *Catostomus commersoni* (White Sucker): molecular analysis predicts distinct precursors for two CRFs and one Urotensin I peptide. *Molecular Marine Biology and Biotechnology*, **1**, 48–57.

Nakano, T. & Tomlinson, N. (1967). Catecholamine and carbohydrate concentrations in rainbow trout *(Salmo gairdneri)* in relation to physical disturbance. *Journal of the Fisheries Research Board of Canada*, **24**, 1701–15.

Nilsson, G.E. (1990). Long-term anoxia in crucian carp: changes in the levels of amino acid and monoamine neurotransmitters in the brain, catecholamines in the chromaffin tissue, and liver glycogen. *Journal of Experimental Biology*, **150**, 295–320.

Nilsson, G.E. & Block, M. (1991). Decreased norepinephrine and epinephrine contents in the chromaffin tissue of rainbow trout *(Oncorhynchus mykiss)* exposed to diethyldithiocarbamate and amylxanthate. *Comparative Biochemistry and Physiology*, **98C**, 391–4.

Nilsson, S. (1984). Adrenergic control systems in fish. *Marine Biology Letters*, **5**, 127–46.

Okawara, Y., Morley, S.O., Burzio, L.O., Ziviers, H., Lederis, K & Richter, D. (1988). Cloning and sequence analysis of cDNA for corticotropin-releasing factor precursor from the teleost fish *Catostomus commersoni*. *Proceedings of the National Academy of Science of the USA*, **85**, 8439–43.

Olivereau, M. & Olivereau, J. (1988). Localization of CRF-like immunoreactivity in the brain and pituitary of teleost fish. *Peptides*, **9**, 13–21.

Olivereau, M. & Olivereau, J. (1990). Effect of pharmacological adrenalectomy on corticotropin-releasing factor (CRF)-like and arginine vasotocin (AVT) immunoreactivities in the brain and pituitary of the eel: immunocytochemical study. *General and Comparative Endocrinology*, **80**, 199–215.

Ortiz, M. & Lutz, P.L. (1995). Brain neurotransmitter changes associated with exercise and stress in a teleost fish (*Sciaenops ocellatus*). *Journal of Fish Biology*, **46**, 551–62.

Pankhurst, N.W. & Dedual, M. (1994). Effects of capture and recovery on plasma levels of cortisol, lactate and gonadal steroids in a natural population of rainbow trout. *Journal of Fish Biology*, **45**, 1013–25.

Perry, S.F. & Reid, S.D. (1992). Relationship between blood O_2 content and catecholamine levels during hypoxia in rainbow trout and American eel. *American Journal of Physiology*, **263**, R240–9.

Perry, S.F. & Reid, S.D. (1993). Beta-adrenergic signal-transduction in fish: interactive effects of catecholamines and cortisol. *Fish Physiology and Biochemistry*, **11**, 195–203.

Perry, S.F. & Reid, S.D. (1994). The effects of acclimation temperature on the dynamics of catecholamine release during acute hypoxia in the rainbow trout. *Journal of Experimental Biology*, **186**, 289–307.

Perry, S.F., Kinkead, R., Gallagher, P. & Randall, D.J. (1989). Evidence that hypoxemia promotes catecholamine release during hypercapnic acidosis in rainbow trout *(Salmo gairdneri)*. *Respiratory Physiology*, **77**, 351–64.

Perry, S.F., Fritsche, R., Kinkead, R. & Nilsson, S. (1991). Control of catecholamine release *in vivo* and *in situ* in the Atlantic cod *(Gadus morhua)* during hypoxia. *Journal of Experimental Biology*, **155**, 540–66.

Pickering, A.D. (1990). Stress and the suppression of somatic growth in teleost fish. In *Progress in Comparative Endocrinology*. Epple, A., Scanes, C.G. & Stetson, M.H. (eds.) pp. 473–9. Wiley–Liss, New York.

Pickering, A.D. & Pottinger, T.G. (1983). Seasonal and diel changes in plasma cortisol levels of the brown trout, *Salmo trutta L. General and Comparative Endocrinology*, **49**, 232–9.

Pickering, A.D. & Pottinger, T.G. (1985). Acclimation of the brown trout, *Salmo trutta L.*, to the stress of daily exposure to malachite green. *Aquaculture*, **44**, 145–52.

Pickering, A.D. & Pottinger, T.G. (1987). Poor water quality suppresses the cortisol response of salmonid fish to handling and confinement. *Journal of Fish Biology*, **30**, 363–74.

Pickering, A.D. & Pottinger, T.G. (1989). Stress responses and disease-resistance in salmonid fish : effects of chronic elevation of plasma cortisol. *Fish Physiology and Biochemistry*, **7**, 253–8.

Pickering, A.D. & Stewart, A. (1984). Acclimation of the interrenal tissue of the brown trout, *Salmo trutta L.*, to chronic crowding stress. *Journal of Fish Biology*, **24**, 731–40.

Pickering, A.D., Pottinger, T.G. & Sumpter, J.P. (1986). Independence of the pituitary–interrenal axis and melanotroph activity in the brown trout, *Salmo trutta L.*, under conditions of environmental stress. *General and Comparative Endocrinology*, **64**, 206–11.

Pickering, A.D., Pottinger, T.G. & Sumpter, J.P. (1987*a*). On the use of dexamethasone to block the pituitary-interrenal axis in the brown trout *Salmo trutta L. General and Comparative Endocrinology*, **65**, 346–53.

Pickering, A.D., Pottinger, T.G., Carragher, J.F. & Sumpter, J.P. (1987*b*). The effects of acute and chronic stress on the levels of reproductive hormones in the plasma of mature male brown trout, *Salmo trutta L. General and Comparative Endocrinology*, **68**, 249–59.

Pickering, A.D., Pottinger, T.G., Sumpter, J.P., Carragher, J.F. & Le Bail, P.Y. (1991). The effects of acute and chronic stress on the levels of circulating growth hormone in the rainbow trout, *Oncorhynchus mykiss. General and Comparative Endocrinology*, **83**, 86–93.

Plotsky, P.M., Cunningham, E.T. & Widmaier, E.P. (1989). Catecholaminergic modulation of corticotropin-releasing factor and adrenocorticotropin secretion. *Endocrine Reviews*, **10**, 437–58.

Pottinger, T.G. & Moran, T.A. (1993). Differences in plasma cortisol and cortisone dynamics during stress in two strains of rainbow trout *(Oncorhynchus mykiss)*. *Journal of Fish Biology*, **43**, 121–30.

Pottinger, T.G., Prunet, P. & Pickering, A.D. (1992). The effects of confinement stress on circulating prolactin levels in rainbow trout *(Oncorhynchus mykiss)* in fresh water. *General and Comparative Endocrinology*, **88**, 454–60.

Pottinger, T.G., Knudsen, F.R. & Wilson, J. (1994). Stress-induced changes in the affinity and abundance of cytosolic cortisol-binding sites in the liver of rainbow trout, *Oncorhynchus mykiss* (Walbaum), are not accompanied by changes in measurable nuclear binding. *Fish Physiology and Biochemistry*, **12**, 499–511.

Pottinger, T.G., Balm, P.H.M. & Pickering, A.D. (1995). Sexual maturity modifies the responsiveness of the pituitary–interrenal axis to stress in male rainbow trout. *General and Comparative Endocrinology*, **98**, 311–20.

Rance, T.A. & Baker, B.I. (1981). The *in vitro* response of the trout interrenal to various fragments of ACTH. *General and Comparative Endocrinology*, **45**, 497–503.

Randall, D.J. & Perry, S.F. (1992). Catecholamines. In *Fish Physiology*, Volume 12. *The Cardiovascular System*. Randall, D.J. & Hoar, W.S. (eds.). Academic Press, New York.

Rand-Weaver, M., Pottinger, T.G. & Sumpter, J.P. (1993). Plasma somatolactin concentrations in salmonid fish are elevated by stress. *Journal of Endocrinology*, **138**, 509–15.

Reid, S.G., Furimsky, M. & Perry, S.F. (1994). The effects of repeated physical stress or fasting on catecholamine storage and release in the rainbow trout, *Oncorhynchus mykiss*. *Journal of Fish Biology*, **45**, 365–78.

Rodrigues, K.T. & Sumpter, J.P. (1983). The distribution of some proopiomelanocortin-related peptides in the pituitary gland of the rainbow trout *Salmo gairdneri*. *General and Comparative Endocrinology*, **51**, 454–59.

Salbert, G., Chauveau, I., Bonnec, G., Valotaire, Y. & Jego, P. (1992). One of the two trout proopiomelanocortin messenger RNAs potentially encodes new peptides. *Molecular Endocrinology*, **6**, 1605–13.

Shrimpton, J.M. & Randall, D.J. (1994). Downregulation of corticosteroid receptors in the gills of coho salmon due to stress and cortisol treatment. *American Journal of Physiology*, **267**, R432–8.

Soma, G.-I., Kitahara, N., Nishizawa, T., Nanami, H., Kotake, C., Okazaki, H. & Andoh, T. (1984). Nucleotide sequence of a cloned cDNA for proopiomelanocortin precursor of chum salmon. *Oncorhynchus keta*. *Nucleic Acids Research*, **12**, 8029–41.

Sumpter, J.P., Pickering, A.D. & Pottinger, T.G. (1985). Stress-induced elevation of plasma α-MSH and endorphin in brown trout *Salmo trutta* L. *General and Comparative Endocrinology*, **59**, 257–65.

Sumpter, J.P., Dye, H.M. & Benfey, T.J. (1986). The effects of stress on plasma ACTH, α-MSH, and cortisol levels in salmonid fishes. *General and Comparative Endocrinology*, **62**, 377–85.

Sumpter, J.P., Carragher, J.F., Pottinger, T.G. & Pickering, A.D. (1987). The interaction of stress and reproduction in trout. In: *Reproductive Physiology of Fish 1987*. Idler, D.R., Crim, L.W. & Walsh, J.M. (eds.) pp. 299–302. Fish Symp. 1987, Sheffield.

Sumpter, J.P., Pottinger, T.G., Rand-Weaver, M. & Campbell, P.M. (1994). The wide-ranging effects of stress on fish. In *Perspectives in Comparative Endocrinology*. Davey, K.G., Peter, R.E. & Tobe, S.S. (eds.) pp. 535–538. National Research Council of Canada, Ottawa.

Takahashi, A., Ogasawara, T., Kawauchi, H. & Hirano, T. (1991). Effects of stress and fasting on plasma growth hormone levels in the immature rainbow trout. *Bulletin of the Japanese Society of Scientific Fisheries*, **57**, 231–5.

Tang, Y. & Boutilier, R.G. (1988). Correlation between catecholamine release and degree of acidosis stress in trout. *American Journal of Physiology*, **255**, R395–9.

Tran, T.N., Fryer, J.N., Lederis, K. & Vaudry, H. (1990). CRF, Urotensin I, and sauvagine stimulate the release of POMC-derived peptides from goldfish neurointermediate lobe cells. *General and Comparative Endocrinology*, **78**, 351–60.

Wagner, G.F. & McKeown, B.A. (1986). Development of a salmon growth hormone radioimmunoassay. *General and Comparative Endocrinology*, **62**, 452–8.

Waugh, D., Anderson, G., Armour, K.J., Balment, R.J., Hazon, N. & Conlon, J.M. (1995). A peptide from the caudal neurosecretory system of the dogfish *Scyliorhinus canicula* that is structurally related to Urotensin I. *General and Comparative Endocrinology*, **99**, 333–9.

White, A. & Fletcher, T.C. (1985). The influence of hormone and inflammatory agents on C-reactive protein, cortisol and alanine aminotransferase in the plaice (*Pleuronectes platessa* L.). *Comparative Biochemistry and Physiology*, **80C**, 99–104.

Wood, C.M., Walsh, P.J., Thomas, S. & Perry, S.F. (1990). Control of red blood cell metabolism in rainbow trout (*Oncorhynchus mykiss*) after exhaustive exercise. *Journal of Experimental Biology*, **154**, 491–507.

Yulis, C.R. & Lederis, K. (1987). Co-localization of the immunoreactivities of corticotropin-releasing factor and arginine vasotocin in the brain and pituitary system of the teleost *Catostomus commersoni*. *Cell and Tissue Research*, **247**, 267–273.

Yulis, C., Lederis, K., Wong, K.-L. & Fisher, A.W.F. (1986). Localization of Urotensin I and corticotropin-releasing factor-like immunoreactivity in the central nervous system of *Catostomus commersoni*. *Peptides*, **7**, 79–86.

G. McDONALD and L. MILLIGAN

Ionic, osmotic and acid–base regulation in stress

Introduction

The focus of this chapter is on the specific causes and consequences of electrolyte and acid–base disturbances provoked by stress in freshwater fish and the mechanisms employed to recover from those disturbances. For this discussion we feel it useful to draw a distinction between 'simple' and 'compound' responses to stress. We define 'simple' responses as being limited to those physiological disturbances directly provoked by the stress hormones, namely adrenaline and cortisol. A 'simple' response most likely occurs when the stressor does not directly injure the fish and water quality remains good (well oxygenated water at circumneutral pH and optimum temperature, with a low ammonia concentration and devoid of toxicants). Many of the stresses of normal aquaculture practice such as routine handling, netting, crowding, confinement and transport (in approximate order of increasing duration), probably evoke a 'simple' stress response only.

'Compound' responses to stress result when one or more additional physiological disturbances compound those evoked by adrenaline and cortisol. Such disturbances can be evoked by stressors that stimulate vigorous activity, by exposure to suboptimal water quality, or by actual physical injury. Vigorous activity usually provokes anaerobiosis and leads to extracellular acid–base disturbances through the release of lactic acid from white muscle. Suboptimal water quality can produce physiological disturbances through direct deleterious effects on the gills or internal organs. The water quality parameters most likely to vary in hatchery practice and to affect ion and/or acid–base balance are oxygen, hardness (i.e. Ca^{2+}, Mg^{2+}), pH, ammonia and salt (i.e. NaCl) levels. Skin abrasions, scale loss and puncture wounds can all be expected to accentuate disturbances, and to prolong the period of stress.

The 'simple' response to stress

Electrolyte disturbances

The emphasis here will be on freshwater salmonids because their responses to stress are well defined. Furthermore, studies of salmonids comprise about two-thirds of the total number of studies of freshwater fish species (both cultured and wild strains, teleosts and non-teleosts), as reported in the extensive review by Barton and Iwama (1991). Other freshwater fish species reared in aquaculture are discussed briefly below (p. 137).

The stress response of freshwater salmonids has been well characterized in numerous studies. While there is considerable variability in the amplitude of the response (as discussed below) its general features are as follows: Adrenaline elevation occurs virtually immediately; it can be detected within seconds and reaches peak values within a minute or two (Mazeaud & Mazeaud, 1981; Randall & Perry, 1992). Cortisol is slower to increase, being significantly elevated in a minute or so, but may continue to increase for up to 4 h before it reaches a peak. Electrolyte losses to the water are similar to the increase in adrenaline, in that they increase virtually immediately to maximum values, and then decline slowly (Fig. 1A). Correlated with the rise in plasma cortisol is an elevation of glucose and a disturbance of plasma electrolyte (i.e. NaCl) balance (Fig. 3). The maximum amplitude of the various changes, relative to resting levels, are approximately as follows (see McDonald & Milligan, 1992): 1000-fold increase in adrenaline, 200-fold increase in cortisol, 3-fold increase in plasma glucose, 40-fold increase in the rate of diffusive loss of Na^+ and Cl^-, and a corresponding 35% loss of whole-body NaCl. As a rule, the degree of change in each of these variables is such that they tend to be correlated with one another, so that each may serve as a measure of the amplitude of the stress response.

While correlated with one another, cause and effect relationships between the stress hormones and the secondary stress indicators are less clearly established. It is generally accepted that the disturbance in blood Na^+ and Cl^- concentrations, at least in simple stress, is caused by the increased levels of adrenaline in the blood (Mazeaud & Mazeaud, 1981). Adrenaline has a myriad of effects but those which are relevant to producing an ionic/osmotic disturbance are: an increase in aortic blood pressure resulting from vasoconstriction and increased cardiac output (Mazeaud & Mazeaud, 1981); an increase in gill diffusing capacity due to increased perfusion of lamellae and an increase in the

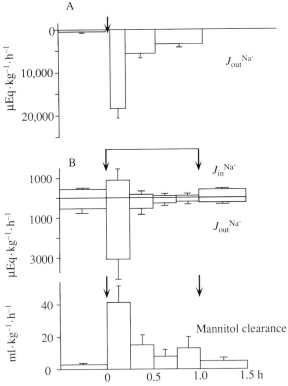

Fig. 1. Effect of adrenaline on branchial Na$^+$ fluxes (J^{Na^+}) in rainbow trout (*Oncorhynchus mykiss*). (A) Adrenaline administered by single intra-peritoneal injection (0.25 μmol/g) at arrow. Mass = 2.2 ± 0.2 g. $n = 8$, temperature, $T = 15.5$ °C. Unpublished data of G. McDonald and V. Thomas. (B) Adrenaline administered by intra-arterial infusion over 1 h (arrows), 10^{-4} M at 10 μl/min. Mass = 311 ± 32 g, $n = 4$, $T = 14$ °C. Mannitol clearance was evaluated from blood levels of [^{3}H]mannitol and [^{3}H]mannitol loss to water. (Redrawn from McDonald & Rogano, 1986).

number of lamellae perfused (Booth, 1979; Randall & Perry, 1992); an increased influx of water across the gills (Adedire & Oduleye, 1984; McDonald, Cavdek & Ellis, 1991); an increased diffusional efflux of electrolytes across the gills (McDonald & Rogano, 1986); an increased urine flow rate (diuresis) (Wood & Randall, 1973; Vermette & Perry, 1987); and an increased urinary loss of electrolytes (Vermette & Perry, 1987).

A decline in plasma ions, therefore, could arise from one or more of the following: increased ion losses across the gills; increased ion losses in the urine; and haemodilution due to net influx of water. While undoubtedly all of these phenomena contribute to the ionic/ osmotic disturbance, increased branchial efflux is, in our experience, by far the largest contributor in 'simple' stress responses (but see discussion on compound responses later). Urinary electrolyte losses are typically less than 10% of branchial diffusional losses under resting conditions, and proportionally even less under stress, despite increased urine flow, because of the relatively large increase in branchial losses (McDonald & Wood, 1981; McDonald, 1983; McDonald, Walker & Wilkes, 1983). Furthermore, the magnitude of the increased diffusional water influx during stress (McDonald et al., 1991) is comparable to the diuresis (Oduleye, 1975), suggesting that there is simply an increased throughput of water during stress, and consequently little net gain of water.

Our recent work (e.g. Gonzalez & McDonald, 1992; McDonald & Robinson, 1993; McDonald, Goldstein & Mitton, 1993; Postlethwaite & McDonald, 1995) has focused on salt disturbances in juvenile salmonids (underyearlings and yearlings) in response to simple stress, induced by either acute or chronic procedures. The acute procedures were brief handling by itself, or in combination with intra-peritoneal injection of either physiological saline or adrenaline solution. In these instances electrolyte disturbances were evaluated by measuring Na^+ losses to the water. The chronic procedures involved net confinement, similar to that employed in previous stress studies (e.g. Strange, Schreck & Ewing, 1978; Davis & Parker, 1986) in which fish were confined in close physical contact for up to 8 h. The net was submerged in a large volume of well-aerated water held at a temperature to which fish had been acclimated (i.e. conditions designed to prevent change or deterioration of water quality as a stress variable). Salt disturbances were assessed from changes in NaCl levels in plasma, in whole bodies or from net losses to the water and, occasionally, all three.

With all of these stress procedures there is typically an initial period of struggling, but this is of short duration (<10 s) and rarely of sufficient vigour to lead to an elevation in the level of lactic acid in muscle or blood. Indeed, in our experience, routine handling of salmonids in well-aerated water produces a stress response without provoking a lactacidosis, i.e. a simple stress response.

In circumstances where the stress response has been produced by the acute administration of adrenaline, the effects on Na^+ and Cl^- diffusion are immediate and substantial; with as much as a 40-fold increase in rates relative to that measured under non-stressful conditions (Fig. 1A). These elevated rates are not sustained as there is a very rapid decline in efflux following the initial stimulation (Fig. 1A). In large part this can be attributed to the very short biological half-life of an injected dose of adrenaline (estimated to be less than 10 min in rainbow trout; Nekvasil & Olson, 1986). However, the reduction in efflux also occurs with continuous infusion of adrenaline (Fig. 1B), suggesting that other factors come into play in diminishing the responsiveness to adrenaline. One such mechanism is the down-regulation of adrenergic receptors known to occur in higher vertebrates (Lefkowitz, Hausdorff & Caron, 1990) but as yet not clearly demonstrated to occur in fish (Randall & Perry, 1992; Gamperl,Wilkinson & Boutilier, 1994). This rapid adaptation of NaCl efflux in response to adrenaline suggests that it is the initial severity of the stress that determines the degree of osmotic disturbance rather than the duration of that stress. This is particularly important in the case of long-distance transport, where reduction of the initial handling stress could be effective in reducing the overall stressful effect.

The specific mechanism by which adrenaline causes the increased efflux of electrolytes is still uncertain but there is strong circumstantial evidence suggesting that it is an effect of increased blood pressure on the gills. Gonzalez and McDonald (1992) concluded, based on an analysis of oxygen uptake and ion loss in rainbow trout in response to adrenaline infusion and other stressful treatments, that the increase in ionic efflux was much larger than could be explained simply on the basis of an increase in functional surface area alone. Instead, they proposed that the important change at the gills is the increase in transmural, and therefore intra-lamellar, pressure. This increased pressure distorts and stretches the paracellular tight junctions and decreases their resistance to ion diffusion, greatly increasing the paracellular permeability of the gills, causing, in effect, ultrafiltration of electrolytes through paracellular channels. Increased branchial clearance of mannitol, considered to be an indicator of paracellular permeability (Kirschner, 1980), during adrenaline infusion (Fig. 1B) supports the notion of an increased leak of electrolytes via the paracellular path.

Role of cortisol in electrolyte disturbances

There is no evidence of which we are aware to indicate that cortisol, by itself, causes electrolyte or osmotic disturbances. However, its release does appear to be responsive to, and proportional to, the ion loss, at least within species and strains. This phenomenon is illustrated in Fig. 2, which shows the results of three separate experiments using juvenile lake trout, *Salvelinus namaycush*, where the degree and duration of confinement were the same (same degree of crowding for 8 h, and thus presumably the same amount of adrenaline release) but where NaCl loss was manipulated by altering the external NaCl concentration. The greater the depression in plasma Na^+ concentration, the greater the elevation of cortisol. This suggests that net ion loss may be stressful in itself, leading to an amplification of the primary stress response, i.e. an elevation in the release of cortisol and adrenaline. The manner in which this feedback may be mediated is unknown, it could be via 'osmoreceptors' in the hypothalamus/pituitary, as has been proposed for prolactin, or via the action of other hormones known to effect cortisol secretion through actions on corticotrophin-releasing hormone (CRH) and adrenocorticotrophic hormone (ACTH) (see chapter by J.P. Sumpter, this volume, for details).

Factors influencing electrolyte losses during stress

Genotype

There are well-documented genetic differences, both amongst species and amongst strains within species, in stress responses as assessed from the elevation in the level of cortisol (e.g. Wedemeyer, 1976; Wydoski,

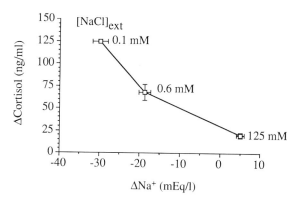

Fig. 2. Changes in plasma Na^+ (ΔNa^+) and cortisol (ΔCortisol) relative to initial (control) values in juvenile lake trout, *Salvelinus namaycush* (30 ± 3 g, 10 °C) after 8 h of net confinement in water containing 0.1, 0.6 or 125 mM NaCl. Data relating to 0.1 mM and 0.6 mM NaCl from McDonald and Robinson (1993), to 125 mM NaCl from G. McDonald and J. Robinson, unpublished.

Wedermeyer & Nelson, 1976; Davis & Parker, 1986; Woodward & Strange, 1987; Pickering, Pottinger & Carragher, 1989, chapter by A.D. Pickering & T.G. Pottinger, present volume). An example of such variations is shown in Fig. 3 for *Salvelinus* sp. (from McDonald *et al.*, 1993). Here, a standardized net confinement stress was employed for all strains (as described previously) so that any variations in sensitivity to stress would be revealed. Note, in particular, the accentuated cortisol response in splake, *S. fontinalis* × *S. namaycush* (Fig. 3A)

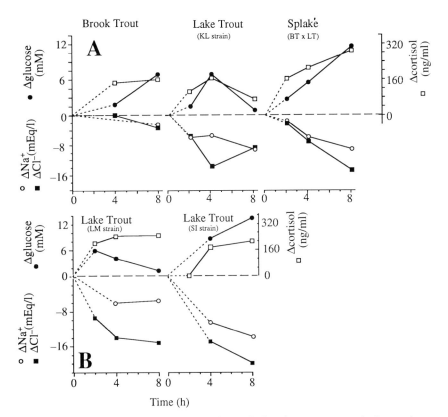

Fig. 3. Intra- and inter-specific variation in responses of plasma ions (Na^+ and Cl^-) and cortisol and glucose [changes relative to initial (control) values] to 8 h of net confinement in juvenile *Salvelinus* sp. at 6–11 °C. (Graph redrawn from McDonald *et al.*, 1993.) (A) Inter-specific comparison: brook trout, *S. fontinalis* (23 g); lake trout, *S. namaycush* [29 g, Killala Lake (KL) strain]; splake, *S. fontinalis* × *S. namaycush* (53 g). (B) Comparison of lake trout strains, Lake Manitou (LM, 25 g) and Slate Island (SI, 22 g) strain.

compared to its two parent species (i.e. brook trout and lake trout). This hybrid 'vigour' is also expressed as a much larger body size at the same age (53 g *versus* 26 g at 18 months) and as an accentuated elevation of glucose concentration and depression of electrolyte concentrations (virtually none in brook trout *versus* 16 mEq/l in splake, Fig. 3A). However, amongst lake trout strains (Fig. 3A, B) there was no clear-cut relationship between cortisol elevation and electrolyte loss. For example, Slate Island strain lake trout showed a larger electrolyte loss compared with the Lake Manitou strain but a lower and slower cortisol elevation. This is perhaps not surprising since ion losses are mediated via adrenaline rather than via cortisol, and there is no reason to presume that cortisol and adrenaline levels would necessarily co-vary.

The implication is that there are genetic differences in either the nature and magnitude of adrenaline release and/or the sensitivity of target tissues to adrenaline which underly these different electrolyte responses. At this point though, there is very little known. Within individual species, there are seasonal and diurnal variations in adrenergic responses (both in the nature of catecholamine mobilization in response to stress and the responsiveness of target tissues to catecholamines; Randall & Perry, 1992), but variations amongst strains and species are not well characterized. Studies of this type are vastly complicated by the challenge of measuring adrenaline levels in blood, as levels increase almost immediately due to any disturbance (Mazeaud, Mazeaud & Donaldson, 1977). This has led to the now widely held view (e.g. Tang & Boutilier, 1988) that plasma catecholamine levels can only accurately be measured by indwelling catheters, implanted with the fish under anaesthesia, and sampled only after the animals have been allowed to recover for at least 48 h from the procedure. Under these circumstances typical resting values for adrenaline are in the range of 1–10 nM (Woodward, 1982; Ristori & Laurent, 1985; Perry *et al.*, 1989). In contrast, acute sampling procedures (cardiac or caudal puncture, etc.) from uncannulated fish lead to much higher 'resting' values. For example, we recently obtained values of >100 nM in juvenile rainbow trout (57 g) sampled by caudal severance, even though the procedure took only 30 s to complete and included an initial 15 s period of MS-222 anaesthesia designed to minimize struggling prior to sampling (M.D. Goldstein & D.G. McDonald, unpublished data).

The virtual impossibility of measuring resting adrenaline levels in fish smaller than 100 g (the practical lower limit for cannulation) and the physiological effects of cannulation itself (see McDonald & Milligan, 1992) mean that we may never know with any certainty the nature of the genetic variability in adrenergic responses. Furthermore, there are some issues that may remain unresolved, such as how cortisol and adrenaline release are related, and whether selection for a reduced cortisol response to stress (Fevolden & Roed, 1993; Pottinger, Moran & Morgan, 1994) has a similar effect on adrenaline release or sensitivity.

Water chemistry

One of the most critical water chemistry variables for stress management in aquaculture is hardness. Natural soft waters tend to be acidic (pH 6–7), and low in Ca^{2+} and NaCl. All three of these conditions are suboptimal for the regulation of ion balance unless the fish have evolved and adapted to these conditions. Low pH and low external NaCl limit Na^+ and Cl^- uptake, and low Ca^{2+} increases the permeability of the gills. Acute exposure of salmonids reared in hard water to soft water is stressful in itself, causing cortisol elevation for several days and persistent electrolyte disturbances requiring up to several weeks to reach a new steady-state (McDonald & Rogano, 1986; McDonald &

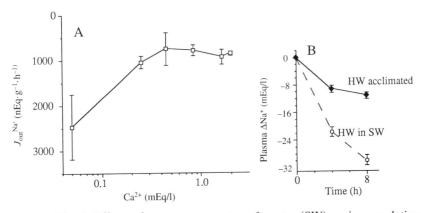

Fig. 4. Effects of acute exposure to soft water (SW) on ion regulation in trout adapted to hard water (HW, NaCl $\approx$ 0.6 mM, Ca^{2+} = 1.0 mM). (A) Diffusive Na^+ loss (J_{out}^{Na+}) over 1 h in relation to external Ca^{2+} for rainbow trout (5–8 g, n = 4 per data point) at 17 °C in 'Na$^+$-free' water (Na^+ < 0.1 mEq/l) where uptake of Na^+ (J_{in}^{Na+}) less than 150 nEq $\cdot$ g^{-1} $\cdot$ h^{-1}. Unpublished data of G. McDonald and J. Lee. (B) Effect of confinement stress in HW and SW (Ca^{2+} = 0.05 mM) on plasma Na^+ in lake trout (29 g, 9 °C, n = 10–44 per data point). Fish were acclimated to hard water then either held in hard water (HW acclimated) or transferred to soft water (HW in SW). Graph redrawn from McDonald and Robinson (1993).

Robinson, 1993). The acute effect of water Ca^{2+} reduction on gill Na^+ permeability of rainbow trout is shown in Fig. 4A. Note the log effect of Ca^{2+} on permeability with a threshold around 0.3 mEq/l, i.e. a hardness of 30 mg/l $CaCO_3$. Diffusional Na^+ losses are about five fold higher in extreme soft water compared to normal hard water (≥ 100 mg/l $CaCO_3$). Not surprisingly, the impact of soft water on plasma Na^+ loss is more severe when animals are acutely stressed in soft water (Fig. 4B). The depression of plasma Na^+ in net-confined fish was about four-fold greater in soft water compared to hard water.

Body size

Net ion losses to the environment are a function of both the permeability of the body surface and the total surface area. Because small fish have a higher gill surface area to mass ratio than larger fish, one could anticipate that small fish would experience greater osmotic disturbances in response to stress. We recently explored the relationship

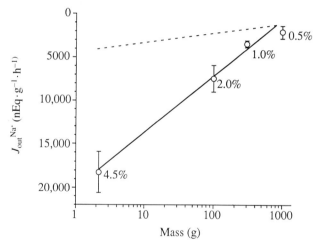

Fig. 5. Effect of acute stress on diffusive Na^+ loss ($J_{out}^{Na^+}$) in rainbow trout at 14 °C in relation to body mass. A standard acute stress was administered by intra-peritoneal injection of adrenaline (0.25 μmol/g). After injection, fish were transferred to flux chambers containing 'Na$^+$-free' hard water, values reported are for the first 10 min after transfer. Values by data points are the estimated losses of Na^+ over first 10 min post-injection as a percentage of total whole body Na^+. The dotted line is the predicted loss based on surface area to volume ratio considerations alone. Unpublished data of G. McDonald and V. Thomas.

between body size and Na^+ loss in rainbow trout (Fig. 5) in response to stress. The stress was standardized by using a single bolus injection of adrenaline (0.25 μmol/g). Ion losses increased with decreasing body mass, but the ion losses were much larger than would be predicted on the basis of surface area to volume ratio alone (upper dotted line in Fig. 5). Clearly, there is some additional scaling phenomenon at work in the gills. Possibilities include size-related changes in tight junction permeability, numbers of cells (tight junctions) per unit area or in adrenergic sensitivity (i.e. nature, location and density of adrenergic receptors).

Recovery from electrolyte disturbances

In comparison to what is known concerning the nature, causes and consequences of stress, relatively little is known of the mechanisms by which fish recover from stress. We have recently examined this issue in detail in the juvenile rainbow trout (Postlethwaite & McDonald, 1995), and found that full recovery in trout was characterized by the return of body NaCl to pre-stress levels, and that Na^+ and Cl^- were recovered over similar time courses. Perhaps not surprisingly, the speed of recovery was dependent upon the degree of the initial disturbance although, generally, recovery was complete within 24–48 h. The mechanism for recovery, at least in the juvenile rainbow trout, comprised a rapid (within 2 h of the start of stress) and progressive (rising eventually to about three fold above control levels) increase in Na^+ and Cl^- uptake by the gills. At the end of the stress period, there was a progressive reduction in diffusive efflux of Na^+ and Cl^-, with efflux reduced to nearly zero. Based on an analysis of changes in uptake kinetics *in vivo* we concluded that ion influx increased because of an activation of inactive transport sites in the gills, whereas efflux was reduced by a reduction in branchial ion permeability achieved without compromising gas exchange. Furthermore, the nature of the adjustments was such as to suggest that both were mediated hormonally. However, we have as yet been unable to identify the specific hormones involved, particularly for the control of permeability, although we showed that a single bolus injection of mammalian growth hormone was able to produce an increase in Na^+ and Cl^- uptake of similar rate and magnitude (Postlethwaite & McDonald, 1995).

Mortality associated with the simple stress response

Much of routine aquaculture stress has probably little, if any, long-term consequences on ionic status, with most fish recovering normal ionic balance in 1–2 days, if not sooner. However, a number of studies of both salmonid and other species have shown that stressors that evoke a simple stress response can be sufficiently severe to cause mortalities (e.g. Strange & Schreck, 1978; Pickering & Pottinger, 1989). Mortality usually starts more than 2 h after the termination of the stress and can continue for several hours to several days. For example, in 1-g rainbow trout exposed to severe confinement stress for 4 h, mortalities began at 8 h post-stress and continued until 40 h post-stress (G. McDonald & P. Prunet, unpublished data). The specific cause(s) of death is rarely certain. For a simple stress response, the proximate cause, especially for those animals dying within a few hours of stress, is probably ion loss. The threshold for mortality occurs when whole-body NaCl loss exceeds approximately 30% of body content, and the effect is, by analogy with our earlier work on electrolyte disturbances in fish exposed to a low pH (Milligan & Wood, 1982), fluid volume disturbances, cardiovascular collapse and oxygen transport failure leading to death. Our recent work provides an illustration of this ion loss threshold (Fig. 6). In 10-g trout, 4 h of confinement stress produced 31% NaCl loss and no mortalities, whereas in 1-g trout the same stress produced 37% NaCl losses and 29% mortality. Achieving this degree of ion loss probably requires at least 30 min of severe 'simple' stress, even for the most stress-sensitive organisms (e.g. chinook salmon) (Strange &

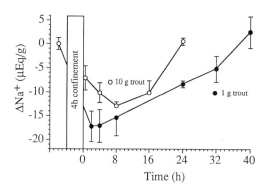

Fig. 6. Changes in whole-body Na^+ (ΔNa^+) in response to 4 h of net confinement and recovery in normal hard water at 12 °C for rainbow trout of two different body sizes (10 ± 2 g, $n = 6$ per time point, and 1.1 ± 0.1 g, $n = 8$ per time point). Data for 10-g fish from Postleth-waite and McDonald (1995), data for 1-g fish from G. McDonald and P. Prunet, unpublished.

Schreck, 1978) under the most stressful circumstances (e.g. small fish in soft water).

Under less stressful circumstances where mortalities are delayed for more than 12–24 h, chronic cortisol elevation becomes increasingly likely as the cause of death (Pickering & Pottinger, 1989). The mechanism of action of cortisol in this instance is probably immunosuppression, which then permits fatal opportunistic infections to develop.

A well-known and effective strategy for reducing mortalities during stressful practices such as handling and transport is the addition of NaCl to the water, either to prevent NaCl losses during stress or aid in recovery of NaCl balance. Iso-osmotic NaCl (i.e. 0.9% NaCl in fresh water) is apparently the procedure most effective at reducing mortality, more effective than $CaCl_2$ (to increase water hardness), anaesthesia, cold temperatures or hyperoxia (e.g. Nikinmaa *et al.*, 1983; Redding & Schreck, 1983; Mazik & Simco, 1994; D.G. McDonald & J.G. Robinson, unpublished observations). The effectiveness of this procedure reinforces the view that NaCl loss is the primary cause of mortality in fish experiencing a 'simple' stress response. Moreover, salt elevation in the water may also be effective in minimizing chronic cortisol elevation (see Fig. 2) thereby reducing the incidence of opportunistic disease.

Compound response to stress: a 'simple' response plus lactacidosis

Angling is perhaps the best example of a circumstance where a 'simple' stress response is combined with lactacidosis to produce a compound response (Booth *et al.*, 1995). This type of response is best modelled experimentally by manually chasing salmonids until they are exhausted, usually for a 5- to 10-min period. This protocol produces similar, though more consistent, results to angling (Beggs, Holeton & Crossman, 1980; Tufts *et al.*, 1991; Ferguson & Tufts, 1992; Booth *et al.*, 1995) and has also been used as a reproducible method for assessing anaerobic capacity in salmonids (e.g. Pearson, Spriet & Stevens, 1990; Wood, 1991; Ferguson, Kieffer & Tufts, 1993). Exhaustive exercise is fuelled primarily by the white muscle mass. Consequently, exhaustion is associated with decreases in concentration of white muscle glycogen, ATP (adenosine triphosphate), PCr (phosphocreatine), pH and increases in lactate and ammonia (Wang, Heigenhauser & Wood, 1994). As with simple stress, there are increases in the plasma levels of 'stress' hormones. Adrenaline and noradrenaline increase immediately after exercise, by about 30-fold and tenfold, respectively, and generally return to pre-exercise levels by 2 h post-exercise. Cortisol is slow to increase; often it is not until 0.5–1 h after exercise that its level is increased significantly to about three times the pre-exercise level, but then it remains elevated for about 4 h (Milligan, 1996). The lactacidosis can be substantial and may be associated with delayed mortality (see below; Wood, Turner & Graham, 1983).

Electrolyte and acid–base disturbances

Whenever a significant lactacidosis occurs, a quite different pattern of plasma electrolyte disturbance emerges (Holeton, Neumann & Heisler, 1983; Wood et al., 1983) from that seen in a 'simple' response. The osmotic disturbances are more complex, for in addition to increased branchial permeability to ions and water, mediated by adrenaline, elevated lactate concentrations in muscle increase muscle intracellular osmotic pressure, leading to a net shift of fluid from the extracellular to the intracellular compartment (Milligan & Wood, 1986). The bulk of the lactate produced (85–90%) is retained within the muscle, with only a small fraction (10–15%) entering the blood (Milligan & Wood, 1986). Consequently, the osmotic gradient persists for a few hours after the stress. The impact of the transcellular osmotic gradient on blood is to cause haemoconcentration, which is manifested as increases in haematocrit, plasma protein and osmolarity, despite net electrolyte losses to the water. A particularly well-documented example of this phenomenon is given by Holeton et al. (1983) where rainbow trout (ca. 900 g) were stimulated to exhaustion by 5 min of mild electrical shock (Fig. 7). Note the initial increase in plasma Na^+ and Cl^-, the subsequent decline in Cl^- (compare to Fig. 3), which is correlated with the rise in plasma lactate, and the marked elevation in haematocrit indicating haemoconcentration. There were only relatively small changes in transbranchial ion fluxes; there was a net Cl^- loss of about 750 nEq $\cdot$ g^{-1} $\cdot$ h^{-1} over the first 1.5 h post-exercise and an equivalent net Na^+ uptake (contrast with Fig. 5). The transbranchial ion fluxes associated with 'compound' stress are clearly different from those seen with 'simple' stress, both in magnitude and direction. Furthermore, similarly to simple stress, the magnitude as well as the nature of the branchial ion losses are size dependent. For example, in smaller fish (e.g. 18–20 g) there are nearly equimolar net Na^+ and Cl^- losses (Fig. 8, left), but in larger fish (250–300 g) there is a greater net branchial loss of Cl^- compared to Na^+ (e.g. Fig. 8, right).

The acidosis associated with 'angling' stress is primarily of metabolic

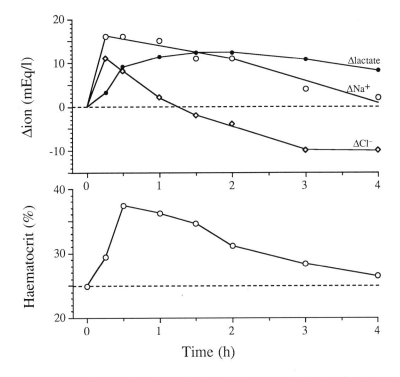

Fig. 7. Plasma chemistry (top) and haematocrit (bottom) of rainbow trout as a function of time during recovery from strenuous muscle activity (evoked by 5 min of mild electric shock). Figure redrawn from Holeton *et al.* (1983). Data are means, $n = 15$, average mass = 935 g, $T = 15\ °C$.

origin (i.e. 'lactic acid' production; Fig. 9A), with a small respiratory component (i.e. elevated CO_2). Blood pH tends to be maximally depressed immediately after cessation of activity and recovers within 4 h (Fig. 9A). Typically, blood lactate levels reach a peak at about 2–4 h after exercise, at levels in excess of 10 mEq/l, and return to resting levels by about 6–8 h (Wang *et al.*, 1994). Exposure to air (60 s) following a bout of vigorous activity leads to a greater elevation in blood lactate and depression in blood pH (Ferguson & Tufts, 1992; Fig. 9A). The latter is due to an accumulation of CO_2, which is rapidly cleared once fish are returned to the water, such that pH is restored by about 4 h. Blood lactate, however, continues to increase throughout the recovery period, indicative of further anaerobiosis

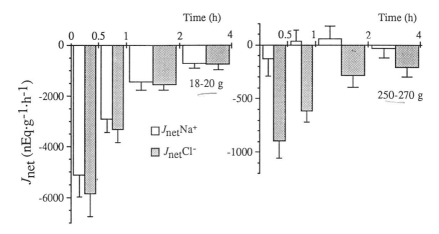

Fig. 8. Net fluxes of Na$^+$ and Cl$^-$ (means $\pm$ 1 SEM) in rainbow trout of two different sizes (19 $\pm$ 3 g, $n = 5$; 263 $\pm$ 11 g, $n = 6$) following 5 min of exhaustive exercise (by manual chasing). Fish were recovered in individual 'flux' chambers (10 ml/g body mass) in normal hard water (Ca$^+$ = 1.0 mM; 100 mg/l CaCO$_3$) and $\approx$0.6 mM NaCl) at 12 °C, and fluxes were calculated from changes in [NaCl] of the water. Unpublished data of J. Kieffer and G. McDonald.

Fig. 9. (A) Blood lactate and pH in rainbow trout (300–500 g) prior to and following 10 min of exhaustive exercise (by manual chasing; striped bar). Fish were transferred to their holding boxes after exercise either directly (i.e. without exposure to air) or following 60 s of exposure to air. Graph redrawn from Ferguson and Tufts (1992). Data are means, $n = 8$ for control and 7 for exposure to air; $T = 8$–10°C. (B) Plasma cortisol, blood lactate and blood pH in rainbow trout (150–250 g) prior to and following 5 min of exhaustive exercise (manual chasing, striped bar). Control: control fish ($n = 11$) were exercised and allowed to recover in still water; metyrapone: fish ($n = 8$) were treated with 3 mg metyrapone/100 g in 0.9% NaCl 1 h prior to exercise and allowed to recover in still water (data from Pagnotta et al., 1994); 1.0 body length · s^{-1} fish ($n = 11$) were allowed to recover from exercise while swimming at 1.0 body length · s^{-1} in a Vogel-type swim tunnel. Means 1 $\pm$ SEM. Unpublished data of B. Hooke and L. Milligan.

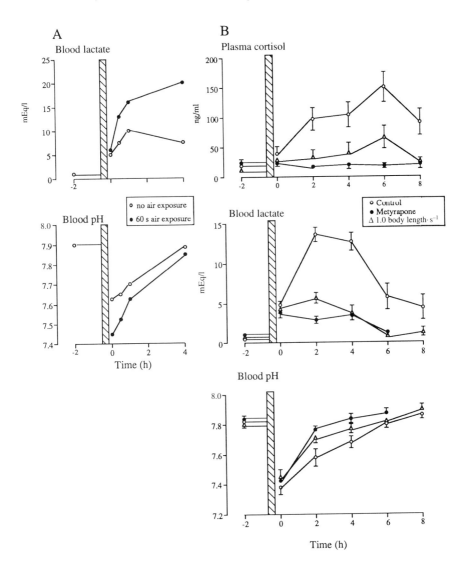

A

Blood lactate

Blood pH

○ no air exposure
● 60 s air exposure

Time (h)

B

Plasma cortisol

○ Control
● Metyrapone
△ 1.0 body length·s⁻¹

Blood lactate

Blood pH

Time (h)

(Ferguson & Tufts, 1992) and may be responsible for increasing mortality (see next section).

The amount of time required for recovery of the lactacidosis appears to be related to cortisol elevation. Typically, cortisol is elevated from about 2–6 h post-exercise (Fig. 9B) and full recovery of metabolite and acid–base status can take up to 12 h (Milligan & Wood, 1986). However, if the rise in cortisol is prevented by treating the fish prior to exercise with metyrapone (which blocks cortisol synthesis), then the time required for recovery of both the blood and muscle disturbances is reduced considerably; from 12 h to about 2 h (Fig. 9B; Pagnotta, Brooks & Milligan, 1994). Interestingly, if fish are allowed to swim at about 1.0 body length/s during recovery from a bout of exhaustive exercise, as opposed to being held in still water, plasma cortisol does not increase and the time for recovery of blood and muscle metabolite and acid–base status is reduced to about 2 h (Fig. 9B; B. Hooke & L. Milligan, unpublished data). This observation suggests two things: (1) it reinforces the notion that elevation of plasma cortisol may be detrimental to recovery from exhaustive exercise; and (2) the stress response associated with exhaustive exercise may not be due to the exercise itself, but rather the lack of activity following exercise. While these observations clearly point to a role, albeit a negative one, for cortisol in regulating recovery from exercise stress, the mechanism by which cortisol is exerting its effect is not clear.

Consequences of a compound stress response

Compound stressors, such as enforced vigorous physical activity, are often associated with *delayed* mortality; death typically occurs 4–12 h after cessation of activity in rainbow trout (Wood *et al.*, 1983). Exposure to air after exercise, even for brief periods (30 s), significantly increases the mortality rate and blood lactate concentration (Fig. 9A). In the 12 h following exercise, Ferguson and Tufts (1992) reported a 12% mortality in rainbow trout (300–500 g), but 30 s of exposure to air immediately following exercise increased mortality to 38%, and 60 s of exposure to air further increased mortality to 72%. The cause of the delayed mortality is uncertain, though Wood *et al.* (1983) suggest that severe intracellular lactacidosis may be a prime factor, since the disturbances to acid–base and electrolyte balance in the blood were not sufficiently severe in themselves to cause death.

Larger fish (200–500 g) may be more prone to post-exercise mortality compared with smaller fish (10–20 g) because larger fish produce more lactate for a given bout of vigorous activity (e.g. 5 min of chasing), and thus more of an acid–base disturbance than do smaller fish (Ferguson *et al.*, 1993). This scaling phenomenon may be related to the relative contributions of aerobic versus anaerobic metabolism to energy production during a bout of vigorous activity. Goolish (1991) suggests that, for a given power output, larger fish rely more heavily upon anaerobic metabolism (hence more lactic acid production) compared with smaller fish.

Interestingly, Booth *et al.* (1995) saw no mortalities in wild, adult (>1 kg) Atlantic salmon, *Salmo salar*, angled to exhaustion (defined by loss of equilibrium), even though there was a brief (*ca.* 5 s) period of exposure to air following angling. It is not clear if the differences in mortality between angled Atlantic salmon and exhaustively exercised rainbow trout represent true species differences, size differences, differences in the exercise protocol, or inherent differences between hatchery (rainbow trout) and wild (Atlantic salmon) fish. What can be said with some certainty, however, is that preventing or minimizing exposure to air after a bout of vigorous physical activity, particularly in larger individuals, will go a long way towards increasing their survival rate.

Stress responses in non-salmonid aquaculture species

Most of the other freshwater fish species that are cultured in North America are warm-water species (channel catfish, largemouth bass, walleye, blue gill, grass carp, striped bass; see Carmichael & Tomasso, 1988). Within this group, the range in sensitivity to stress is much greater than in salmonids (c.f. Davis & Parker, 1986). Nonetheless, the stress responses are qualitatively similar in nature.

Particular attention has focused recently on stress responses of striped bass, *Morone saxatilis*, as this species appears especially sensitive to stress. In a procedure producing a 'simple' stress response [5 h of transport (at 18 kg/100 l) in soft water, 28 mg/l $CaCO_3$], plasma Na^+ and Cl^- fell by 25 and 35 mEq/l respectively, and cortisol rose to 600 ng/ml (Mazik, Simco & Parker, 1991), i.e. substantially greater than in *Salvelinus* sp. (Fig. 3). After transport, the fish were transferred to hard water holding tanks for recovery. In this condition, plasma Na^+ and Cl^- continued to fall and cortisol to rise for at least a further 6 h (Fig. 10A). Subsequently, fish began to die, and mortality eventually reached 100% by 4 weeks post-transport. This is a more extreme response than is typical for salmonids, particularly with respect to delayed mortalities.

Furthermore, a compound stress response is apparently much more readily evoked in striped bass compared to salmonids (Young & Cech,

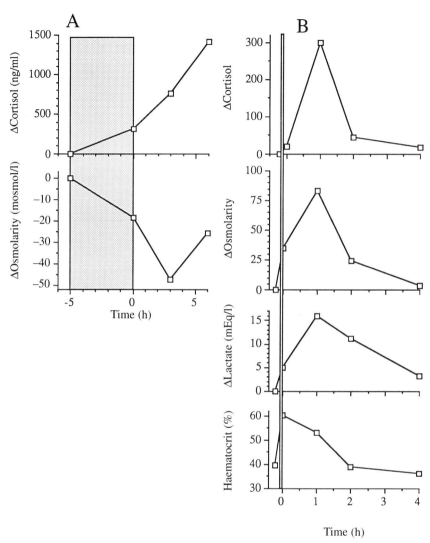

Fig. 10. Responses to stress in striped bass (*Morone saxitilis*). (A) Responses to 5 h hauling in soft water (28 mg/l CaCO₃) at a density of 18 kg/100 l at 12 °C. Data are means (*n* = 10), average mass = 72 g. ΔOsmolarity was calculated from the sum of Na⁺, Cl⁻, glucose, Ca²⁺ and K⁺ using the correction of 0.91 × [Na + Cl] of Robertson, 1989. Data are from Mazik *et al.* (1991). (B) Capture, net confinement and crowding in a shallow pan of water for 90 s followed by recovery in 1.0% NaCl at 22 °C. Data are means (*n* = 9–12), average mass = 24 g. Data of Young and Cech (1993).

1993). After only 90 s of a stress procedure comprised of capture, net confinement and crowding in a shallow pan of water, blood lactate rose by 18 mEq/l (a near maximal lactate response), cortisol rose by 300 ng/ml, osmolarity by about 50 mosmol/l and haematocrit by about 50% (Fig. 10B, compare to Fig. 7 for rainbow trout).

On almost the other end of scale, largemouth bass, *Micropterus salmoides*, showed a much smaller response to a similar treatment. Thirty minutes of net confinement produced almost no change in plasma osmolarity and an increase of cortisol of only 85 ng/ml. Continuous stress for 48 h was required to produce disturbances comparable to those experienced by striped bass after only 5 h of transport. Furthermore, largemouth bass could tolerate up to 12 h of confinement with no resulting mortalities (Carmichael *et al.*, 1984*a, b*).

Summary and conclusions

Simple and compound stress responses are physiologically very different events with different consequences for the fish. Simple stress responses, the type most likely encountered in routine aquaculture practice, result in net ion losses at the gills, the severity of which depend on such factors as genotype, body size and water quality. Mortality associated with simple stress responses appears to be a function of the extent of ion loss, which is correlated with body size; larger fish suffer less ion loss and thus lower mortality rates. Compound stress responses are less likely to be encountered in routine aquaculture, except perhaps by striped bass, but are of more serious consequence. The physiological disturbances are complicated by the production of lactic acid, also scaling with body size, so that larger fish may be more susceptible to delayed mortality than smaller fish. Cortisol elevation associated with a compound stress response is detrimental to recovery from that stress. Furthermore, cortisol elevation with a simple stress response is reinforced by ion loss. Therefore, preventing or minimizing the elevation of cortisol by, for example, placing recovering fish in an elevated salt concentration or in a current and allowing them to swim may assist the recovery process and reduce stress-associated mortalities.

The available data strongly suggest that there are scaling effects which may alter the physiological response to stress, and should be taken into consideration when assessing stress-amelioration methods. Clearly, there is still much to learn about the physiology of the stress response in fish. Specifically there is a need to broaden the scope of study to the non-salmonid species used in aquaculture, and to focus more fully on the technically more challenging task of characterizing adrenergic responses to stress.

References

Adedire, C.O. & Oduleye, S.O. (1984). Stress-induced water permeability changes in the tropical cichlid, *Oreochromis niloticus* (Trewavas). *Journal of Fish Biology*, **25**, 463–71.

Barton, B.A. & Iwama, G.K. (1991). Physiological changes in fish from stress in aquaculture with emphasis on the response and effects of corticosteroids. *Annual Review of Fish Diseases*, **1**, 3–26.

Beggs, G.L., Holeton, G.F. & Crossman, E.J. (1980). Some physiological consequences of angling stress in muskellunge, *Esox masquinongy* Mitchill. *Journal of Fish Biology*, **17**, 649–59.

Booth, J.L. (1979). The effects of oxygen supply, epinephrine and acetylcholine on the distribution of blood flow in trout gills. *Journal of Experimental Biology*, **83**, 31–9.

Booth, R.K., Kieffer, J.D., Davidson, K., Bielak, A.T. & Tufts, B.L. (1995). Effects of late-season catch and release angling on anaerobic metabolism, acid–base status, survival & gamete viability in wild Atlantic salmon (*Salmo salar*). *Canadian Journal of Fisheries and Aquatic Sciences*, **52**, 283–90.

Carmichael, G.J. & Tomasso, J.R., (1988). Survey of fish transportation equipment and techniques. *Progressive Fish-Culturist*, **50**, 155–9.

Carmichael, G.J., Tomasso, J.R., Simco B.A. & Davis, K.B. (1984*a*). Characterization and alleviation of stress associated with hauling largemouth bass. *Transactions of the American Fisheries Society*, **113**, 778–85.

Carmichael, G.J., Tomasso, J.R. Simco, B.A. & Davis, K.B. (1984*b*). Confinement and water quality-induced stress in largemouth bass. *Transactions of the American Fisheries Society*, **113**, 767–77.

Davis, K.B. & Parker, N.C. (1986). Plasma corticosteroid stress response of fourteen species of warm water fish to transportation. *Transactions of the American Fisheries Society*, **115**, 495–9.

Ferguson, R.A. & Tufts, B.L. (1992). Physiological effects of brief air exposure in exhaustively exercised rainbow trout (*Oncorhynchus mykiss*): implications for 'catch and release' fisheries. *Canadian Journal of Fisheries and Aquatic Sciences*, **49**, 1157–62.

Ferguson, R.A., Kieffer, J.D. & Tufts, B.L. (1993). The effects of body size on the acid–base and metabolite status in the white muscle of rainbow trout before and after exhaustive exercise. *Journal of Experimental Biology*, **180**, 195–207.

Fevolden, S.E. & Roed, K. (1993). Cortisol and immune characteristics in rainbow trout (*Oncorhynchus mykiss*) selected for high or low tolerance to stress. *Journal of Fish Biology*, **43**, 919–30.

Gamperl, A.K., Wilkinson, M. & Boutilier, R.G. (1994). β-Adrenoreceptors in the trout (*Oncorhynchus mykiss*) heart: characterization, quantification & effects of repeated catecholamine exposure. *General and Comparative Endocrinology*, **95**, 259–72.

Gonzalez, R.J. & McDonald, D.G. (1992). The relationship between oxygen consumption and ion loss in a freshwater fish. *Journal of Experimental Biology*, **163**, 317–32.

Goolish, E.M. (1991). Aerobic and anaerobic scaling in fish. *Biological Reviews*, **66**, 33–56.

Holeton, G.F., Neumann, P. & Heisler, N. (1983). Branchial ion exchange and acid–base regulation after strenuous exercise in rainbow trout (*Salmo gairdneri*). *Respiration Physiology*, **51**, 303–18.

Kirschner, L.B. (1980). Uses and limitations of inulin and mannitol for monitoring gill permeability changes in the rainbow trout. *Journal of Experimental Biology*, **85**, 203–17.

Lefkowitz, R.J., Hausdorff, W.P. & Caron, M.G. (1990). Turning off the signal-desensitization of beta-adrenergic function. *FASEB J*, **4**, 2881–9.

Mazeaud, M. & Mazeaud, F. (1981). Adrenergic responses to stress in fish. In *Stress and Fish*. Pickering, A.D. (ed.) pp. 49–75. Academic Press, New York.

Mazeaud, M.M., Mazeaud, F. & Donaldson, E.M. (1977). Primary and secondary effects of stress in fish: some new data with a general review. *Transactions of the American Fisheries Society*, **106**, 201–12.

Mazik, P.M. & Simco, B.A. (1994). Effects of size, water hardness, salt levels and MS-222 on the survival and stress response in striped bass (*Morone saxitilis*). In *High Performance Fish*. MacKinlay, D.D. (ed.) pp. 425–30. American Fisheries Society, Vancouver.

Mazik, P.M., Simco, B.A. & Parker, N.C. (1991). Influence of water hardness and salts on survival and physiological characteristics of striped bass during and after transport. *Transactions of the American Fisheries Society*, **120**, 121–6.

McDonald, D.G. (1983). The interaction of environmental calcium and low pH on the physiology of the rainbow trout, *Salmo gairdneri*. I. Branchial and renal net ion and H^+ fluxes. *Journal of Experimental Biology*, **102**, 123–40.

McDonald, D.G. & Milligan, C.L. (1992). Chemical properties of the blood. In *Fish Physiology*, Volume XIIB. Hoar, W.S., Randall, D.J. & Farrell, A.P. (eds.) pp. 55–133. Academic Press, New York.

McDonald, D.G. & Robinson, J.G. (1993). Handling and confinement stress in the lake trout (*Salvelinus namaycush*) – effects of water hardness and genotype. *Transactions of the American Fisheries Society*, **122**, 1146–55.

McDonald, D.G. & Rogano, M.S. (1986). Ion regulation by the rainbow trout, *Salmo gairdneri*, in ion-poor water. *Physiological Zoology*, **59**, 318–31.

McDonald, D.G. & Wood, C.M. (1981). Branchial, and renal acid and ion fluxes in the rainbow trout, *Salmo gairdneri* at low environmental pH. *Journal of Experimental Biology*, **93**, 101–18.

McDonald, D.G., Walker, R.L. & Wilkes, P.R.H. (1983). The interaction of calcium and low pH on the physiology of the rainbow trout, *Salmo gairdneri* II. Branchial iono-regulatory mechanisms. *Journal of Experimental Biology*, **102**, 141–55.

McDonald, D.G., Cavdek, V. & Ellis, R. (1991). Gill design in freshwater fishes: interrelationships amongst gas exchange, ion regulation & acid–base regulation. *Physiological Zoology*, **64**, 103–23.

McDonald, D.G., Goldstein, M.D. & Mitton, C. (1993). Responses of hatchery-reared brook trout, lake trout & splake to transport stress. *Transactions of the American Fisheries Society*, **122**, 1127–38.

Milligan, C.L. (1996). Metabolic recovery from exhaustive exercise in rainbow trout. *Comparative Biochemistry and Physiology*, **113A**, 51–60.

Milligan, C.L. & Wood, C.M. (1982). Disturbances in haematology, fluid volume distribution and circulatory function associated with low environmental pH in the rainbow trout, *Salmo gairdneri*. *Journal of Experimental Biology*, **99**, 397–416.

Milligan, C.L. & Wood, C.M. (1986). Intracellular and extracellular acid–base status and H^+ exchange with the environment after exhaustive exercise in the rainbow trout. *Journal of Experimental Biology*, **123**, 93–121.

Nekvasil, N.P. & Olson, K.R. (1986). Plasma clearance, metabolism & tissue accumulation of ^{3}H-labelled catecholamines in trout. *American Journal of Physiology*, **250**, R519–R525.

Nikinmaa, M., Soivio, A., Nakari, T. & Lindgren, S. (1983). Hauling stress in brown trout (*Salmo trutta*): physiological responses to transport in fresh water or salt water, and recovery in natural brackish water. *Aquaculture*, **34**, 93–9.

Oduleye, S.A. (1975). The effects of calcium on water balance of the brown trout *Salmo trutta*. *Journal of Experimental Biology*, **63**, 343–56.

Pagnotta, A., Brooks, L. & Milligan, L. (1994). The potential regulatory roles of cortisol in recovery from exhaustive exercise in rainbow trout. *Canadian Journal of Zoology*, **72**, 2136–46.

Pearson, M.P., Spriet, L.L. & Stevens, E.D. (1990). Effect of sprint training on swim performance and white muscle metabolism during exercise and recovery in rainbow trout. *Journal of Experimental Biology*, **149**, 45–60.

Perry, S.F., Kinkead, R., Gallaugher, P. & Randall, D.J. (1989). Evidence that hypoxemia promotes catecholamine release during hypercapnic acidosis in rainbow trout (*Salmo gairdneri*). *Respiration Physiology*, **77**, 351–64.

Pickering, A.D. & Pottinger, T.G. (1989). Stress responses and disease resistance in salmonid fish: effects of chronic elevation of plasma cortisol. *Fish Physiology and Biochemistry*, **7**, 253–8.

Pickering, A.D., Pottinger, T.G. & Carragher, J.F. (1989). Differences in the sensitivity of brown trout, *Salmo trutta* L. and rainbow trout, *Salmo gairdneri* Richardson, to physiological doses of cortisol. *Journal of Fish Biology*, **34**, 757–68.

Postlethwaite, E.K. & McDonald, D.G. (1995). Mechanisms of Na^+ and Cl^- regulation in freshwater-adapted rainbow trout (*Oncorhynchus mykiss*) during exercise and stress. *Journal of Experimental Biology*, **198**, 295–304.

Pottinger, T.G., Moran, T.A. & Morgan, J.A.W. (1994). Primary and secondary indices of stress in the progeny of rainbow trout (*Oncorhynchus mykiss*) selected for high and low responsiveness to stress. *Journal of Fish Biology*, **44**, 149–63.

Randall, D.J. & Perry, S.F. (1992). Catecholamines. In *Fish Physiology*, Volume XIIB. (Hoar, W.S. Randall, D.J. & Farrell, A.P. (eds.) pp. 255–300. Academic Press, New York.

Redding, J.M. & Schreck, C.B. (1983). Influence of ambient salinity on osmoregulation and cortisol concentration in yearling coho salmon during stress. *Transactions of the American Fisheries Society*, **112**, 800–7.

Ristori, M.T. & Laurent, P. (1985). Plasma catecholamines and glucose during moderate exercise in the trout: comparison with bursts of violent activity. *Experimental Biology*, **44**, 247–53.

Robertson, J.D. (1989). Osmotic constituents of the blood plasma and parietal muscle of *Scyliorhinus canicula* (L). *Comparative Biochemistry and Physiology*, **93A**, 799–805.

Strange, R.J. & Schreck, C.B. (1978). Anesthetic and handling stress on survival and cortisol concentration in yearling chinook salmon (*Oncorhynchus tshawytscha*). *Journal of the Fisheries Research Board of Canada*, **35**, 345–49.

Strange, R.J., Schreck, C.B. & Ewing, R.D. (1978). Cortisol concentrations in confined juvenile chinook salmon (*Oncorhynchus tshawytscha*). *Transactions of the American Fisheries Society*, **107**, 812–19.

Tang, Y. & Boutilier, R.G. (1988). Correlation between catecholamine release and degree of acidotic stress in trout. *American Journal of Physiology*, **255**, R395–R399.

Tufts, B.L., Tang, Y., Tufts, K. & Boutilier, R.G. (1991). Exhaustive exercise in 'wild' Atlantic salmon (*Salmo salar*): acid–base regulation and blood gas transport. *Canadian Journal of Fisheries and Aquatic Sciences*, **48**, 868–74.

Vermette, M.G. & Perry, S.F. (1987). The effects of prolonged epinephrine infusion on the physiology of the rainbow trout, *Salmo gairdneri*. III. Renal ionic fluxes. *Journal of Experimental Biology*, **128**, 269–85.

Wang, Y., Heigenhauser, G.J.F. & Wood, C.M. (1994). Integrated responses to exhaustive exercise and recovery in rainbow trout white muscle: acid–base, phosphogen, carbohydrate, lipid, ammonia, fluid volume and electrolyte metabolism. *Journal of Experimental Biology*, **195**, 227–58.

Wedemeyer, G. (1976). Physiological response of juvenile coho salmon (*Oncorhynchus kisutch*) and rainbow trout (*Salmo gairdneri*) to handling and crowding stress in intensive fish culture. *Journal of the Fisheries Research Board of Canada*, **33**, 2699–702.

Wood, C.M. (1991). Acid–base and ion balance, metabolism and their interactions, after exhaustive exercise in fish. *Journal of Experimental Biology*, **160**, 285–308.

Wood, C.M. & Randall, D.J. (1973). The influence of swimming activity on water balance in the rainbow trout (*Salmo gairdneri*). *Journal of Comparative Physiology*, **82**, 257–76.

Wood, C.M., Turner, J.D. & Graham, M.S. (1983). Why do fish die after severe exercise? *Journal of Fish Biology*, **22**, 189–201.

Woodward, C.C. & Strange, R.J. (1987). Physiological stress responses in wild and hatchery-reared rainbow trout. *Transactions of the American Fisheries Society*, **116**, 574–79.

Woodward, J.J. (1982). Plasma catecholamines in resting rainbow trout, *Salmo gairdneri* Richardson, by high pressure liquid chromatography. *Journal of Fish Biology*, **21**, 429–32.

Wydoski, R.S., Wedemeyer, G.A. & Nelson, N.C. (1976). Physiological response to hooking stress in hatchery and wild rainbow trout (*Salmo gairdneri*). *Transactions of the American Fisheries Society*, **105**, 601–6.

Young, P.S. & Cech, J.J. (1993). Effects of different exercise conditioning on stress responses and recovery in cultured and wild young-of-the-year striped bass (*Morone saxatilis*). *Canadian Journal of Fisheries and Aquatic Sciences*, **50**, 2094–99.

C.B. SCHRECK, B.L. OLLA and M.W. DAVIS

Behavioral responses to stress

Introduction

Alterations in the behavior of fish induced by stress may significantly affect activities essential for survival including food acquisition, predator avoidance and habitat selection. Behavioral measures of stress have proved to be sensitive indicators of the complex consisting of biochemical and physiological changes that occurs in response to stress (for reviews see: Marcucella & Abramson, 1978; Olla, Pearson & Studholme, 1980a; Beitinger, 1990; Schreck, 1990; Scherer, 1992). Additionally, behavioral measures have the distinct advantage over other methodology of being readily interpreted within an ecological context, thereby increasing the efficacy of extrapolating laboratory results to the natural environment. The ecological relevance of findings will be directly dependent upon how closely laboratory-based behavioral norms are derived from natural life habits, behavior and ecological requirements (Olla et al., 1980a; Scherer, 1992).

Changes in behavior caused by stress may, in many cases, be adaptive and thereby increase the probability of survival (Fig. 1). When an animal is exposed to a perturbation, the first line of defense is a behavioral one designed to lessen the probability of death or the metabolic costs incurred by maintaining physiological homeostasis (Olla et al., 1980a). The behavior elicited may be an avoidance or other behavioral change that mitigates exposure. However, if avoidance or behavioral mitigation is not possible, induced changes in behavior may then reflect deleterious changes in how an animal senses and responds to its environment. Significant departures from behavioral norms would be suggestive of a decreased probability of survival.

The initial response to stressful situations is neuroendocrine (Mazeaud, Mazeaud & Donaldson, 1977; Barton & Iwama, 1991). This physiological reaction is initiated only by stimuli that are perceived centrally by the fish and involve the catecholamines and glucocortico-

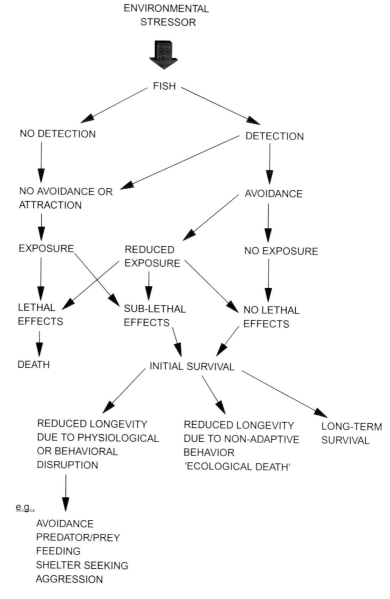

Fig. 1. A flow chart indicating possible behavioral responses to an environmental stressor and their consequences (Modified from Olla *et al.*, 1980*a*).

steroid hormones, cortisol being the primary one in teleostean fishes (Schreck, 1981, 1992; Schreck' & Li, 1991). The neural and hormonal factors initiated by perception of 'potentially life-threatening situations' result in a cascade of other physiological responses that affect tertiary stress reactions such as behavior (Schreck, 1981; Schreck & Li, 1991). The immediate neuronal response to even very acute stresses initiates other physiological and behavioral responses that vary considerably in their recovery rates. Recovery from very brief (i.e., seconds to minutes) stressful experiences may take up to days or weeks for certain physiological systems or behavioral activities (Fig. 2).

Physiology and behavior are intrinsically coupled. As such, physiological mechanisms are responsible for initiating and maintaining

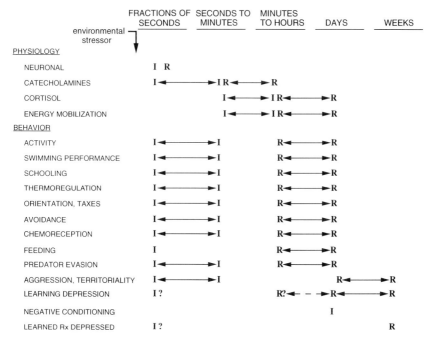

Fig. 2. The time that it takes to initiate relevant physiological or behavioral responses to stressful situations and the time that it takes for these responses to return back to levels under prestress conditions. Information in this figure is based on data presented in the literature reviewed in this chapter. I = initiation; R = recovery; the arrows indicate the variation reported for the response; the dashed arrow and question marks indicate some degree of uncertainty. Rx = reaction or response.

behavioral sequences. This holds for the resting animal, the active animal, and the stressed animal. Physiological responses to stress can be viewed as controlling and directing factors (Fry, 1947) with regard to behavior. They can turn certain behaviors on and others off, or modify the degree to which behaviors are exhibited. Those stressors that are perceived as life threatening, even if they are not, lead to the classical physiological stress response as first described by Selye (1936, 1950). Such stressors are characterized by a direct linkage between the stimulus and the physiological or behavioral reaction. Other environmental factors, for example certain xenobiotics, pathogens and temperature, may result in physiological and behavioral stress responses indirectly; such responses may be modified manifestations of the direct responses. For example, certain toxicants may interfere with normal metabolism which then leads to breakdown of regulatory mechanisms; temperature affects physiological function because of properties associated with the Q_{10} (temperature coefficient) phenomenon; and pathogens may cause tissue damage that leads to abnormal physiological function (Schreck, 1981).

Consequently, the suite of responses that potentially may be elicited from the behavioral repertoire in response to an external stimulus would depend upon the stimulus first being sensed and perceived, either directly or indirectly, as described above, before it could be acted upon (Olla *et al.*, 1980*a,b*). Thus sensory acuity, responsiveness, and the adaptive behaviors elicited in response to those sensory inputs are intimately involved with the success of food acquisition, predator avoidance, orientation and migration, reproduction, and learning. The measurement of sensory capabilities linked to elicitation of behaviors critical for survival represents another relevant avenue for examining behavioral manifestations of stress.

When behavioral measures are made concomitantly with biochemical and physiological parameters, they serve to demonstrate the ecological significance of those other measures. In turn, correlating physiological and biochemical measures with behavioral indices can further understanding of the underlying mechanisms responsible for behavioral alteration.

In what follows we attempt to show, by example, various approaches that have been applied for examining the effects of stress on the behavior of fish. Although our intention is not to present an exhaustive review of the literature, we do cite pertinent studies which have utilized behavior as a tool to measure stress and include some examples of behavioral measures that show potential as tools to assess stress effects (for reviews see: Marcucella & Abramson, 1978; Olla *et al.*, 1980*a*; Beitinger, 1990; Schreck, 1990; Scherer, 1992). We begin by describing factors that are stress inducing and that have been shown to alter behavior, followed by descriptions of the variety of behaviors that have been shown to be affected by stress.

Causitive factors for stress-induced alterations in behavior

A wide variety of physical, chemical, and biological stressors have been identified as causative factors responsible for altering the behavior of fish that potentially are inimical to their survival. Handling of fish associated with capture and confinement in tanks (Pickering, Pottinger & Christie, 1982; Mesa, 1994; Olla, Davis & Schreck, 1995) and cages (Srivastava, Brown & Allen, 1991; Cordero, Beveridge & Muir, 1994), fishing with various gear types (Wardle, 1986; Fernö, 1993; Isaksen & Valdemarsen, 1994), and electrofishing (Mesa & Schreck, 1989) have all been observed to induce levels of stress which altered behavior.

Exposure to chemical factors that induce stress and result in changes in behavior include organochlorines (Henry & Atchison, 1984; Brown *et al.*, 1987; Kruzynski & Birtwell, 1994), heavy metals (Peterson, 1976; Henry & Atchison, 1979*a,b*), asbestos (Belanger *et al.*, 1986), oil (Folmar *et al.*, 1981; Butler, Trivelpiece & Miller, 1982), ammonia (Woltering, Hedtke & Weber, 1978; Hedtke & Norris, 1980), chlorine (Jones & Hara, 1988), acids (Nakamura, 1986; Jones, Brown & Hara, 1987), and anesthetics (Lewis *et al.*, 1985; Belanger *et al.*, 1986; Quinn, Olson & Konecki, 1988; Taylor, 1988). The primary physiological responses recover from acute netting and handling sorts of stresses in about 6 h to 1 day while it may take 10 days to 2 weeks to recover if the stressful situation persists but is not lethal (Schreck, 1981).

The exposure of fish to environmental extremes has caused demonstrable alterations in behavior. For example, subjecting fish to upper and lower temperature extremes resulted in significant departures from established behavioral norms for Atlantic mackerel, *Scomber scombrus* (Olla *et al.*, 1975), bluefish, *Pomatomus saltatrix* (Olla & Studholme, 1971; Olla *et al.*, 1975), tautog, *Tautoga onitis* (Olla & Studholme, 1975; Olla *et al.*, 1975, 1978), and tilapia, *Tilapia mossambica* (Kutty & Sukumaran, 1975). Behavioral thermoselection of preferred temperatures has also been shown to be altered by stress in bluegill, *Lepomis macrochirus* (Beitinger & Magnuson, 1975), Atlantic salmon, *Salmo salar* (Peterson, 1976), and bluefish (Olla, Studholme & Bejda, 1985), as has exposure to salinity gradients for salmonids (Kruzynski & Birtwell, 1994; Kruzynski, Birtwell & Chew, 1994).

Intraspecific behavioral interactions have been observed to elicit stress responses that are manifested as behavioral change when fish that exhibit territoriality or aggressive competitive interactions for food or mates are confined in either a laboratory or hatchery environment. For example, agonistic encounters under captive conditions have been shown to elicit stress-induced behavioral alterations in salmonids (Ejike & Schreck, 1980; Pazkowski & Olla, 1985; Abbott & Dill, 1989; Pottinger & Pickering, 1992), and bluegill (Henry & Atchison, 1986). Such socially induced behavioral responses have obvious ecological implications. Social stress in Nile tilapia, *Oreochromis niloticus*, results in subordinate animals consuming more of their carbohydrate energy reserves which has consequences for growth (deOliveira-Fernandes & Volpato, 1993). Density-dependent social interactions also affect spawning behavior. On crowded spawning grounds individual sockeye salmon, *Oncorhynchus nerka*, that do not hold territories are under stress and may not reproduce (Semenchenko, 1988; Parenskiy, 1989, 1990).

Exposure to high current speed or prolonged exercise has induced stress and altered swimming performance of sockeye salmon (Bams, 1967) and tilapia (Kutty & Sukumaran, 1975). Swimming in shallow water is stressful to goldfish, as evidenced by such fish having elevated circulating levels of cortisol (Fryer, 1975). Wave action can also direct behavior. Rainbow trout in net-pens increased swimming and social interactions as wave frequency and height increased (Srivastava *et al.*, 1991).

Types of stress-induced alterations in behavior

Evaluation of behavioral performance in response to stressors has involved various degrees of complexity. The simplest experiments exposed fish to stress and then measured changes in single behaviors. More broadly encompassing were those experiments which have examined responses of complex behaviors to stressors which involve the integration of more than one simple behavior into ecologically meaningful responses (Scherer, 1992; Kruzynski & Birtwell, 1994). What follows are examples of the range of behaviors that have been affected by, and the behavioral measures that have been used to evaluate the effects of, stress. Furthermore, since behavioral measures of stress have not been widely employed in relation to fish, we provide examples of the potential of employing such measures.

Avoidance, sensory behavior and activity

The first line of defense of a fish to an environmental challenge including stress is avoidance. Removal of, or retreat from, the causative agent lessens or forestalls the physiological and biochemical changes that are associated with stress (Olla *et al.*, 1980a; Hocutt, Denoncourt & Stauffer, 1982; Nakamura, 1986). However, if stress cannot be avoided, as may be the case with anthropogenic stressors, and a fish survives the immediate insult, the complex physiological and biochemical changes that occur may be expressed as alterations in behavior.

For avoidance to be successful, fish must first be capable of sensing a threatening stimulus, recognizing it as inimical and then responding appropriately (Olla *et al.*, 1980*a*). While these steps involve various levels of complexity, each is potentially vulnerable to the effects of stress. Stress can thus serve as a directive factor (Fry, 1947), modifying behavior as the gradient in the severity of stress changes.

Decreases in sensory capabilities can be measured directly by behavioral methodology, with departures from established baselines serving to assess stress effects. For example, established chemoreceptive thresholds for food extracts based on behavioral measures in the dungeness crab, *Cancer magister*, were shown to be altered by exposure to petroleum hydrocarbons (Pearson *et al.*, 1981). The potential applicability of this technique to fish is supported by experiments which have shown the efficacy of establishing behavioral chemosensory thresholds to food extracts in the red hake, *Urophycis chuss* (Pearson, Miller & Olla, 1980), that could readily serve as baselines to quantify the effects of stress. Further support comes from experiments which show a change in the chemosensory threshold exhibited by food-deprived sablefish, *Anoplopoma fimbria* (Løkkeborg *et al.*, 1995). The interconnection of chemosensory ability and key behavioral patterns can make the analysis of stress a complex task involving synergistic effects at several levels of the behavioral response. In these cases, single-factor studies may not allow predictions to be made as to the ecological consequences of stress-induced changes.

Deficits in sensory behavior, stimulus recognition and response capabilities may also play a role in determining whether appropriate responses are elicited to environmental changes that are associated with orientation, taxes and habitat choice. For example, stress has been shown to alter behavioral responses to current for coho salmon, *Oncorhynchus kisutch*, and green sunfish, *Lepomis cyanellus* (Belanger *et al.*, 1986), temperature for Atlantic salmon (Peterson, 1976) and walleye pollock, *Theragra chalcogramma* (Sogard & Olla, 1996), depth preference for killifish, *Fundulus heteroclitus* (Butler *et al.*, 1982) and salmonids (Kruzynski *et al.*, 1994), shelter for Arctic char, *Salvelinus alpinus* (Jones & Hara, 1988) and light for chinook salmon, *Oncorhynchus tshawytscha* (Sigismondi & Weber, 1988) and walleye pollock (Olla & Davis, 1992). Once behavioral responses to these environmental factors are altered, behaviors that are maladaptive for particular environments may be expressed.

Changes in activity have been observed to be a sensitive indicator of stress. For example, sublethal upper and lower extremes in temperature caused marked increases in swimming activity in bluefish (Olla & Studholme, 1971) and Atlantic mackerel (Olla *et al.*, 1975), and decreases in activity and changes in social behavior and shelter seeking in tautog (Olla & Studholme, 1975; Olla *et al.*, 1978). Differences between species in endurance and sustainable swimming rates in fish forced to swim continuously in response to an optimotor stimulus were measured for Atlantic mackerel, herring, *Clupea harengus*, and saithe, *Pollachius virens* (He & Wardle, 1988). When bluegill were exposed to methyl parathion or copper, both spontaneous activity and ventilation rate increased (Henry & Atchison, 1984, 1986). Exposure to chrysotile asbestos caused coho salmon and green sunfish to lose equilibrium and swim chaotically (Belanger *et al.*, 1986). Interestingly, while stress associated with transportation decreased swimming ability in migrating chinook salmon smolts (Maule *et al.*, 1988), holding these fish in captivity for several days, i.e., preventing the migratory behavior, led to physiological stress and subsequent death. Pathogenic microorganisms and other parasites are also known to affect behavior. Symptoms of certain diseases included increased activity and 'flashing', and advanced stages of diseases usually included decreased activity and lethargy (Warren, 1991; Thoesen, 1994). Certain parasites induce leaping and rolling behavior in Atlantic salmon in net-pens (Furevik *et al.*, 1993). Certain pathogens also result in behavioral shifts in the temperature preference of various species. Fish may exhibit behavioral fever where sick individuals select a temperature that is warmer than normal in an effort to fight the infection. Such a febrile response to pathogenic organisms has been noted in bluegill sunfish (Reynolds, 1977), goldfish (Covert & Reynolds, 1977; Reynolds, Covert & Casterlin, 1978), and tilapia (Tsai & Hoh, 1995).

Predator–prey interactions

The ability to avoid or evade predators, as is the case for other stimuli, is based on first being able to sense a predator via any combination of sensory modalities and then responding appropriately. However, the small number of studies that have examined the effects of stress on predator avoidance or evasion have generally not separated out sensory deficits from performance deficits. One of the few studies published on this subject showed that osmotic stress in Atlantic salmon smolts caused reactive and escape distances from predators to be reduced, resulting in an increased probability of predation (Järvi, 1989*a,b*). The potential does exist for separating sensory and performance deficits in modalities other than vision. For example, the detection of predator alarm substance in pearl dace, *Semotilus margarita*, via chemoreception causes rapid and erratic antipredatory swimming behavior followed by prolonged inactivity (Rehnberg, Smith & Sloley, 1987). Stress has the potential to alter these responses and to lead to disorientation and inappropriate responses that may impede the ability of salmonids to avoid or evade predators (Kruzynski *et al.*, 1994).

There is clear evidence that stress causes significant deficits in predator avoidance/evasion. Exposure of guppies, *Poecilia reticulata*, to pentachlorophenol was followed by an increase in the efficiency of capture by largemouth bass, *Micropterus salmoides* (Brown *et al.*, 1985). The imposition of handling stress caused significant increases in the vulnerability to predation for coho smolts (Fig. 2; Olla & Davis, 1989; Olla, Davis & Schreck, 1992) and for both coho and chinook salmon smolts (Olla *et al.*, 1995). Recovery of predator evasion behavior in salmonids varied considerably over time depending on the stock, species and level of stress that was imposed. In some cases levels of cortisol did not reflect the recovery from stress (Fig. 3; Olla *et al.*, 1992), while in other cases there was close agreement (Olla *et al.*, 1995).

Different stocks of fish have been observed to respond to a particular stressor in different ways. These responses may be the result of differences in species, genetic make-up, or rearing history. In performance testing of wild and hatchery-reared sockeye salmon, differences were noted in swimming and predator evasion abilities associated with rearing history (Bams, 1967). Similarly, using a predation assay, differences in the time taken to recover from single and multiple handling stresses were noted to occur between stocks of coho and chinook salmon from different hatcheries (Olla *et al.*, 1995), suggesting that behavioral assays of stress tolerance may be a sensitive method for identifying problems with hatchery rearing techniques. Clearly, results of experiments with specific fish stocks can only be extrapolated to other stocks after careful consideration of the possible effects of relevant intrinsic and extrinsic factors on behavioral response capabilities.

Schooling behavior, which is an adaptive strategy that has evolved in part as an antipredator mechanism, may also be affected by stress and thereby affect vulnerability to predation. For example, pulp mill effluent disrupted schooling behavior of chum salmon, *Oncorhynchus keta*, and coho salmon (Birtwell & Kruzynski, 1989). Cadmium disrupted schooling in fathead minnows, *Pimephales promelas*, by inducing random and chaotic swimming which resulted in increased rates of predation (Sullivan *et al.*, 1978).

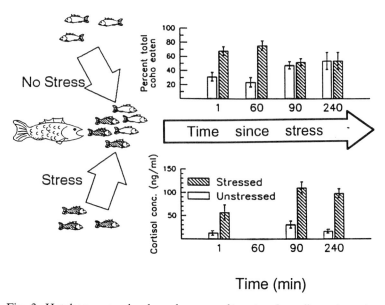

Fig. 3. Hatchery–reared coho salmon smolts eaten by a lingcod predator (percent of the total number eaten in a trial; mean ± SE) and cortisol concentration prior to predation trial (mean ± SE). Stressed fish were held in a bucket without water for 1 min, then allowed to recover for 1, 60, 90 or 240 min. Unstressed fish were allowed to recover from the transfer for 24 h prior to predation trial. (Modified from Olla, Davis & Ryer, 1994).

Innate attraction between fish can operate in concert with, or in opposition to, other environmental factors to control the spatial cohesiveness of a school of fish (Ryer & Olla, 1991). When the predation threat is high, fright acts in conjunction with innate attraction. In contrast, hunger operates in opposition to innate attraction, resulting in decreased attraction and less cohesive schools as fish spread out in search of food, resulting in the reward of foraging being balanced with the risk of predation (Olla et al., 1996). Stress may interrupt any stage of this behavioral decision-making process and lead to alteration of spatial distribution and success of foraging or predator avoidance.

When predator and prey are subjected to stress, the effects may be more complex and difficult to predict. For example, stress had a greater effect on the behavior of the predator, largemouth bass, thus enhancing the survival of prey, mosquitofish, *Gambusia affinis* (Woltering et al., 1978). Alternatively, stress may have manifested a greater effect on prey (chinook salmon), resulting in an increase in predation rate by brook trout, *Salvelinus fontinalis* (Hedtke & Norris, 1980). In the latter case, i.e., of chinook salmon, the synergistic effects of predator presence and toxicant produced increased predation of the prey. In more ecologically realistic experiments, 'ecological death' results from the synergistic effects of multiple stressors, where the fish has not died but is unable to continue behaving in a meaningful way, resulting in eventual mortality from various causes including predation (Henry & Atchison, 1979*a*; Kruzynski & Birtwell, 1994).

Feeding

Searching for and finding food is a more complex behavior than simple activity or predator avoidance. Feeding can require appetite, visual and chemosensory ability, restricted area searching, responding to and capturing prey, and handling and ingestion of prey. Stress has been observed to disrupt any or all of these components of feeding behavior (Beitinger, 1990). For example, food searching behavior in sablefish, as mediated by chemosensory ability, was shown to increase with food deprivation (Løkkeborg *et al.*, 1995). Low pH has been shown to depress attraction to food scent and to depress feeding in Arctic char (Jones *et al.*, 1987) and fathead minnow (Lemly & Smith, 1987). Exposure to petroleum decreased feeding by coho salmon on rainbow trout, *Oncorhynchus mykiss* (Folmar *et al.*, 1981) while handling stress depressed feeding for brown trout, *Salmo trutta* (Pickering *et al.*, 1982), and stress caused by electrofishing decreased feeding of cutthroat trout, *Oncorhynchus clarki* (Mesa & Schreck, 1989).

We have commonly observed that stress caused by handling, such as transferring fish from one tank to another, frequently results in loss of feeding behavior for various periods of time, depending on the severity of the stress and the physiological state of the fish. Resumption of feeding is frequently correlated with a re-establishment of normal physiological status. For example, the time taken for Pacific salmon to resume feeding after stress is roughly the same as the time it takes for physiological indices of stress, such as circulating levels of cortisol or glucose, to return to prestress concentrations.

Shelter seeking

Shelter seeking involves the coordination of several simpler behaviors, any of which can be disrupted by stress. Some species of fish seek shelter in response to a negative phototaxis, a positive thigmotaxis, as a fright response to chemosensory detection of a predator or noxious scent, as a response to a visual fright stimulus, or to thermal stress. For example, shelter seeking was stimulated by low pH for arctic char (Jones *et al.*, 1987), by predator alarm substance for pearl dace (Rehnberg *et al.*, 1987), and by electrofishing stress for cutthroat trout (Mesa & Schreck, 1989). Exposure to the antisapstain fungicide 2-(thiocyanomethylthio) benzothiazole (TCMTB) caused shelter seeking in salmonids to be an inappropriate response, where disoriented fish sought cover in an area of abundant predators, into which the fish would not ordinarily go (Kruzynski *et al.*, 1994). Increased temperature caused tautog to increase shelter seeking and association with shelter, which resulted in decreased activity, feeding, and aggression (Olla & Studholme, 1975).

More severe stress can modify the behavioral response to less severe concurrent stress. Shelter seeking after exposure to light was prolonged (i.e., it took the fish longer to reach cover) in chinook salmon that first received a brief handling stress (Sigismondi & Weber, 1988).

Aggression

Territorial and agonistic behavior can be disrupted by stressors and may result in maladaptive responses including displacement of fish from optimal habitats (Rand, 1985; Beitinger, 1990). Territoriality in many species, whether for defending food resources, potential mates or an optimal habitat, results in the formation of a dominance hierarchy composed of dominant, intermediate and subordinate individuals. Concentrations of neuropeptides associated with the central nervous system, such as dopamine and norepinephrine, that are sensitive to stress are influenced by social rank in rainbow trout (Vosyliene, Petrauskiene & Prekeris, 1993), and these peptides may direct behavior. In juvenile coho salmon, status in a dominance hierarchy was inversely related to a chronic state of stress (Fig. 4; Ejike & Schreck, 1980). This effect was clearly manifested for coho smolts as reduced feeding and the inhibition in subordinates to behave aggressively, even in the absence of any fish that had previously been dominant (Fig. 3; Pazkowski & Olla, 1985). Differential access to food by members of a social hierarchy has been observed to result in growth depensation, an increase in the variance of size or weight in a population, for medaka, *Oryzias latipes* (Magnuson, 1962) and chum salmon (Davis & Olla, 1987), and has higher metabolic costs associated with being a subordinate steelhead trout (Li & Brocksen, 1977; Metcalfe, 1986; Abbott & Dill, 1989). Growth depensation can also lead to higher levels of aggression, stress, and mortality. Excessive aggressiveness may be maladaptive in the

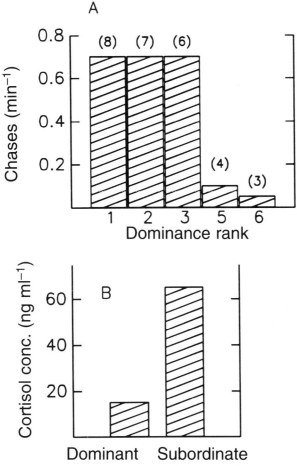

Fig. 4. (A) Number of chases (per minute) by dominant hatchery-reared coho salmon smolts during the disassembly of a group from eight to three fish. (B) Cortisol concentration from dominant and subordinate fish in groups of six hatchery-reared coho. (Modified from Olla *et al.*, 1994).

wild, as fish must balance feeding and other social behaviors against predatory risk (Metcalfe, Huntingford & Thorpe, 1987; Ryer & Olla, 1991).

Animals can display dominance not associated with aggression; they do not have to be overtly aggressive to display their dominance. Such individuals appear physiologically unstressed. Aggressive individuals can also become dominant consequent to their behavior; this type of

individual exhibits physiological signs of stress (Leshner, 1978). Aggressive behaviors may be driven in part by circulating androgen levels that,
in turn, may be stimulated by social stimuli. This mechanism of aggressive control has been suggested for *Haplochromis burtoni* and *Xiphoph-
orus helleri* (Hannes & Franck, 1983).

Stress may result in either increased or decreased aggressive behavior.
For example, exposing bluegills to copper caused aggression to increase
(Henry & Atchison, 1986). In contrast, decreases in aggression were
observed for tautog when subjected to thermal stress (Olla *et al.*, 1985)
and for bluegill when exposed to methyl parathion (Henry & Atchison,
1984). In a social hierarchy, the effects of stress appear to be greatest
for the primary dominant and subordinate fish, while others in the
group appear less affected (Henry & Atchison, 1979*b*).

Effects of stress on early life stages

Little information exists on the impact of stress in the early life stages
of fish despite the fact that most of the natural mortality in fish occurs
during the larval stages, due to starvation and predation (Hunter, 1981;
Bailey & Houde, 1989). While it is known that the magnitude of the
physiological stress response shifts through time as salmonids go through
the parr–smolt transformation (Schreck, 1982; Barton *et al.*, 1985),
nothing is known about the nature of the physiological stress response
during true larval stages. The few extant studies have primarily focused
on the effects of starvation on vulnerability to predation. Larval fish
from the earliest stages of development possess the capability to respond
to environmental factors, the most important of which are associated
with food acquisition and predator evasion (Olla *et al.*, 1996). Their
vulnerability to stress is in part due to limited swimming capabilities
and low energy reserves. Thus simple effects on swimming ability may
have major consequences for survival. When early-stage larvae are
stressed by a lack of food or poor quality of food, they may die directly
from starvation or indirectly from developmental anomalies which preclude their adaptive behavioral responses to environmental challenges
(see, for example, Blaxter & Ehrlich, 1974; Davis & Olla, 1992; Olla &
Davis, 1992). In walleye pollock larvae, starvation caused a depression
and eventual cessation of positive phototaxis (Olla & Davis, 1992)
while nutritionally deficient prey caused abnormalities in gas bladder
development, leading to a lack of buoyancy control and the ability to
maintain the vertical position in a water column and early death
(Davis & Olla, 1992). Inappropriate and nonadaptive behavioral pat-

terns may lead to greater vulnerability to predators and increased mortality (Fuiman & Magurran, 1994).

A wide array of chemical stresses can affect behavior of larval fishes (Birge *et al.*, 1992). While there tend to be interspecific differences, avoidance, feeding, and chemoreaction behavior were generally affected.

Although larvae may be potentially more sensitive to the effects of stress compared with fish at the more mature stages of development, the difficulty of culturing and handling larvae has inhibited research that directly links the effects of stress with behavior.

Effects of stress on learning and conditioning

Given the correct context, teleosts are quite capable of learning (see review by Gleitman & Rozin, 1971). Environmental factors that can influence learning ability include pathogens, toxicants, and seasonality. Catfish with ichthyophthirious had retarded learning. The pesticide DDT [1,1,1-trichloro-2,2-bis(*p*-chlorophenyl)ethane] modified the learning of a simple conditioned response in rainbow trout (McNicholl & Mackay, 1975*a,b*). There were seasonal rhythms in learning ability in goldfish, perhaps associated with reproductive condition (Shashoua, 1973).

Many of the stimuli used in classic conditioning experiments with fish involve stressors of various kinds as negative reinforcement. These include touch, temperature, electric shock, and bright lights (Gleitman & Rozin, 1971).

There is little information directly considering the effects of stress on learning ability or memory. An early study concluded that stress (being taken out of the water) in goldfish interfered with initial acquisition of a learning task but enhanced it if the fish had already acquired a little of the knowledge (Laird, Baud & Hoffman, 1972). The same species trained by avoidance training in a shuttlebox had amnesia if given a 1-day isolation stress prior to training (Laudien *et al.*, 1986). Chemical stressors can also interfere with learning. Goldfish exposed to acid stress had slower operant learning than did controls and also more variable behavior (Garg & Garg, 1992).

Stress can interfere with performance of learned responses as well. Positive conditioning was used to train fully acclimated coho salmon juveniles to recognize water from a hatchery other than the one in which they were being reared. Over many repeated trials, 100% of the fish learned the correct response in a few days. However, if naive fish were transported from one hatchery to another, it required several weeks of recovery before the fish could be taught. Furthermore, in a controlled experiment in which all of the fish had acquired the learned response, fish given a 2-h transportation stress required 48 h of post-stress recovery before all individuals again performed positively in the tests (W.A. Sandoval & C.B. Schreck, unpublished data).

Recovery from physiological stress is important for overall performance of transported fish. Coho salmon that had been transported as juveniles and allowed a few months to recover before release to the ocean had greater survivorship to adulthood than those liberated into streams immediately after transport when they still exhibited elevated cortisol titers (Schreck et al., 1989). It is difficult to discern whether recovery is needed for physical or mental capacity in the fish, or for both. Positive conditioning associating stress (lowering the water level) with reward (food) enhanced the physiological recovery and overall performance of transported chinook salmon juveniles compared to control fish. Psychological 'hardening' was at least partially responsible for these results because the positively conditioned fish also performed better than fish that received the stimulus (lowering of the water) not associated with food prior to transport (Schreck et al., 1995).

Training that presumably involved more severe stress but a more realistic situation was also more advantageous than a less direct method that constituted a negative reinforcement type of learning. The severity of the physiological stress response was moderated and the behavioral predator avoidance response was more appropriate in Atlantic salmon trained to avoid predators by direct rather than indirect prior experience with predators (Järvi & Uglem, 1993).

Stress may have direct effects on the central nervous system of fishes. Corticosteroid receptors that could bind to deoxyribonucleic acid (DNA) have been characterized from the cytosol of juvenile chinook salmon brains (Knoebl, Fitzpatrick & Schreck, 1996), and numbers of these receptors were depressed following 3 h of crowding stress (I. Knoebl, M.S. Fitzpatrick, & C.B. Schreck, unpublished data from author's laboratory). It is interesting to speculate that sustained elevation of cortisol levels, such as during chronic stress, could cause loss of memory and learning dysfunction due to neuronal death in fish, as has been shown by Sapolsky, Krey and McEwen. (1984a,b) to occur in mammals.

Summary and conclusions

The response to stress is manifested in a physiological cascade that initiates with central perception of the situation. Neuronal mediation activates the chromaffin tissue and the hypothalamo–pituitary–interrenal axis. The hormones of stress, catecholamines and cortisol, regulate the secondary and tertiary physiological stress responses including the energetic state of the fish.

The neuronal stress response also mediates a variety of behavioral stress responses that can be modified by the physiological state of the fish. Behaviors that are affected include: avoidance, sensory behavior, and activity; predator–prey interactions; feeding; shelter seeking; aggression; and learning and conditioning. These behaviors are generally modified almost instantaneously with perception of the stress and may require minutes to days or weeks to return to prestress conditions once stress has been removed. Behaviors with a very high survival value tend to return most rapidly.

Larval life history stages may be potentially more sensitive to stress compared with other stages, yet little is known about behavior and stress in fish at this time in their lives. Fish undergoing metamorphosis may be most vulnerable to stress. Nonetheless, with the ever-increasing interest in fish culture, the study of stress and behavior of larval fish appears to be an important and fertile field for the future. Critical, however, is the need for an understanding of how stress affects behavior of larval fish and other life history stages given the potential failure in recruitment in natural populations due to anthropogenic disturbance, such as global climate change, pollution, or habitat alteration.

References

Abbott, J.C. & Dill, L.M. (1989). The relative growth of dominant and subordinate juvenile steelhead trout (*Salmo gairdneri*) fed equal rations. *Behaviour*, **108**, 104–13.

Bailey, K.M., & Houde, E.D. (1989). Predation on eggs and larvae of marine fishes and the recruitment problem. *Advances in Marine Biology*, **25**, 1–83.

Bams, R.A. (1967). Differences in performance of naturally and artificially propagated sockeye salmon migrant fry, as measured with swimming and predation tests. *Journal of the Fisheries Research Board of Canada*, **24**, 1117–53.

Barton, B.A. & Iwama, G. (1991). Physiological changes in fish from stress in aquaculture with emphasis on the response and effects of corticosteroids. *Annual Review of Fish Diseases*, **1**, 3–26.

Barton, B.A., Schreck, C.B., Ewing, R.D., Hemmingsen, A.R. & Patiño, R. (1985). Changes in plasma cortisol during stress and smoltification in coho salmon, *Oncorhynchus kisutch*. *General and Comparative Endocrinology*, **59**, 468–71.

Beitinger, T.L. (1990). Behavioral reactions for the assessment of stress in fishes. *Journal of the Great Lakes Research*, **16**, 495–528.

Beitinger, T.L. & Magnuson, J.J. (1975). Influence of social rank and size on thermoselection behavior of bluegill (*Lepomis macrochirus*). *Journal of the Fisheries Research Board of Canada*, **32**, 2133–6.

Belanger, S.E., Schurr, K., Allen, D.J. & Gohara, A.F. (1986). Effects of chrysotile asbestos on coho salmon and green sunfish: evidence of behavioral and pathological stress. *Environmental Research*, **39**, 74–85.

Birge, W.J., Hoyt, R.D., Black, J.A., Kercher, M.D. & Robison, W.A. (1992). Effects of chemical stresses on behavior of larval and juvenile fishes and amphibians. In *Proceedings of the Larval Fish Conference*, Kingston, Rhode Island.

Blaxter, J.H.S., & Ehrlich, K.F. (1974). Changes in behavior during starvation of herring and plaice larvae. In *The Early Life History of Fish*. Blaxter, J.H.S. (ed.) pp. 575–88. Springer-Verlag, Berlin.

Birtwell, I.K. & Kruzynski, G.M. (1989). *In situ* and laboratory studies on the behavior and survival of Pacific salmon (genus *Oncorhynchus*). *Hydrobiologia*, **188/189**, 543–60.

Brown, J.A., Johansen, P.H., Colgan, P.W. & Mathers, R A. (1985). Changes in the predator-avoidance behaviour of juvenile guppies (*Poecilia reticulata*) exposed to pentachlorophenol. *Canadian Journal of Zoology*, **63**, 2001–5.

Brown, J.A., Johansen, P.H., Colgan, P.W. & Mathers, R.A. (1987). Impairment of early feeding behavior of largemouth bass by pentachlorophenol exposure: a preliminary assessment. *Transactions of the American Fisheries Society*, **116**, 71–8.

Butler, R.G., Trivelpiece, W. & Miller, D.S. (1982). The effects of oil, dispersant, and emulsions on the survival and behavior of an estuarine teleost and an intertidal amphipod. *Environmental Research*, **27**, 266–76.

Cordero, F.J., Beveridge, M.C.M. & Muir, J.F. (1994). A note on the behaviour of adult Atlantic halibut, *Hippoglossus hippoglossus* (L.), in cages. *Aquaculture and Fisheries Management*, **25**, 475–81.

Covert, J.B.& Reynolds, W.W. (1977). Survival value of fever in fish. *Fish Review*, **267**, 43–5.

Davis, M.W. & Olla, B.L. (1987). Aggression and variation in growth of chum salmon (*Oncorhynchus keta*) juveniles in seawater: effects of limited rations. *Canadian Journal of Fisheries and Aquatic Sciences*, **44**, 192–7.

Davis, M.W. & Olla, B.L. (1992). Comparison of growth, behavior and lipid concentrations of walleye pollock *Theragra chalcogramma* larvae fed lipid-enriched, lipid deficient and field collected prey. *Marine Ecology Progress Series*, **90**, 23–30.

Ejike, C. & Schreck, C.B. (1980). Stress and social hierarchy rank in coho salmon. *Transactions of the American Fisheries Society*, **109**, 423–6.

Fernö, A. (1993). Advances in understanding of basic behavior: consequences for fish capture studies. *ICES Marine Science Symposium*, **196**, 5–11.

Folmar, L.C., Craddock, D.R., Blackwell, J.W., Joyce, G. & Hodgins, H.O. (1981). Effects of petroleum exposure on predatory behavior of coho salmon (*Oncorhynchus kisutch*). *Bulletin of Environmental Contamination and Toxicology*, **27**, 458–62.

Fuiman, L.A. & Magurran, A.E. (1994). Development of predator defenses in fishes. *Reviews of Fish Biology and Fisheries*, **4**, 145–83.

Fry, F.E.J. (1947). Effects of the environment on animal activity. *University of Toronto Biology Series* 55, *Publication of the Ontario Fisheries Research Laboratory*, No. 68. University of Toronto Press, Toronto. pp. 1–62.

Fryer, J.N. (1975). Stress and adrenocorticosteroid dynamics in the goldfish, *Carassius auratus. Canadian Journal of Zoology*, **53**, 1012–20.

Furevik, D.M., Bjordal, A., Huse, I. & Fernoe, A. (1993). Surface activity of Atlantic salmon *(Salmo salar* L.) in net pens. *Aquaculture*, **110**, 119–28.

Garg, R. & Garg, A. (1992). Operant learning of goldfish exposed to pH depression in water. *Journal of Environmental Biology*, **13**, 1–6.

Gleitman, H. & Rozin, P. (1971). Learning and memory. In *Fish Physiology*. Randall, D.J. (ed.) pp. 191–278. Academic Press, New York.

Hannes, R.-P. & Franck, D. (1983). The effect of social isolation on androgen and corticosteroid levels in a cichlid fish (*Haplochromis burtoni*) and in swordtails (*Xiphonphorus helleri*). *Hormones and Behaviour*, **17**, 292–301.

He, P. & Wardle, C.S. (1988). Endurance at intermediate swimming speeds of Atlantic mackerel, *Scomber scomberus* L., herring, *Clupea harengus* L., and saith, *Pollachius virens* L. *Journal of Fish Biology*, **33**, 255–66.

Hedtke, J.L. & Norris, L.A. (1980). Effect of ammonium chloride on predatory consumption rates of brook trout (*Salvelinus fontinalis*) on juvenile chinook salmon (*Oncorhynchus tshawytscha*) in laboratory streams. *Bulletin of Environmental Contamination and Toxicology*, **24**, 81–9.

Henry, M.G. & Atchison, G.J. (1979a). Behavioral changes in bluegill (*Lepomis macrochirus*) as indicators of sublethal effects of metals. *Environmental Biology of Fishes*, **4**, 37–42.

Henry, M.G. & Atchison, G.J. (1979*b*). Influence of social rank on the behaviour of bluegill, *Lepomis macrochirus* Rafinesque exposed to sublethal concentrations of cadmium and zinc. *Journal of Fish Biology*, **15**, 309–15.

Henry, M.G. & Atchison, G.J. (1984). Behavioral effects of methyl parathion on social groups of bluegill *(Lepomis macrochirus)*. *Environmental Toxicology and Chemistry*, **3**, 399–408.

Henry, M.G. & Atchison, G.J. (1986). Behavioral changes in social groups of bluegills exposed to copper. *Transactions of the American Fisheries Society*, **115**, 590–5.

Hocutt, C.H., Denoncourt, R.F. & Stauffer, J.R. Jr. (1982). Observations of behavioural responses of fish to environmental stress in situ. *Journal of Applied Ecology*, **19**, 443–51.

Hunter, J.R. (1981). Feeding ecology and predation of marine fish larvae. In *Marine Fish Larvae: Morphology, Ecology, and Relation to Fisheries*. Lasker, R. (ed.) pp. 33–77. Washington Sea Grant Program, University of Washington, Seattle. Distributed by University of Washington Press.

Isaksen, B. & Valdemarsen, J.W. (1994). Bycatch reduction in trawls by utilizing behaviour differences. In *Marine Fish Behaviour in Capture and Abundance Estimation*. Fernö, A. & Olsen, S. (eds.) pp. 69–83. Fishing News Books, London.

Järvi, T. (1989*a*). The effects of osmotic stress on the anti-predator behaviour of Atlantic salmon smolts: a test of the 'maladaptive anti-predator behaviour' hypothesis. *Nordic Journal of Freshwater Research*, **65**, 71–9.

Järvi, T. (1989*b*). Synergistic effect on mortality in Atlantic salmon, *Salmo salar*, smolt caused by osmotic stress and presence of predators. *Environmental Biology of Fishes*, **26**, 149–52.

Järvi, T. & Uglem, I. (1993). Predator training improves the anti predator behavior of hatchery reared Atlantic salmon (*Salmo salar*) smolt. *Nordic Journal of Freshwater Research*, **68**, 63–71.

Jones, K.A. & Hara, T.J. (1988). Behavioral alterations in Arctic char *(Salvelinus alpinus)* briefly exposed to sublethal chlorine levels. *Canadian Journal of Fisheries and Aquatic Sciences*, **45**, 749–53.

Jones, K.A., Brown, S.B. & Hara, T.J. (1987). Behavioral and biochemical studies of onset and recovery from acid stress in Arctic char (*Salvelinus alpinus*). *Canadian Journal of Fisheries and Aquatic Sciences*, **44**, 373–81.

Knoebl, I., Fitzpatrick, M.S. & Schreck, C.B. (1996). Characterization of a glucocorticoid receptor in the brains of chinook salmon, *Oncorhynchus tshawytscha*. *General and Comparative Endocrinology*, **101**, 195–204.

Kruzynski, G.M. & Birtwell, I.K. (1994). A Predation bioassay to quantify the ecological significance of sublethal responses of juvenile chinook salmon (*Oncorhynchus tshawytscha*) to the antisapstain fungicide TCMTB. *Canadian Journal of Fisheries and Aquatic Sciences*, **51**, 1780–90.

Kruzynski, G.M., Birtwell, I.K. & Chew, G.L. (1994). Behavioural approaches to demonstrate the ecological significance of exposure of juvenile Pacific salmon (genus *Oncorhynchus*) to the antisapstain fungicide TCMTB. *Journal of Aquatic Ecosystems and Health*, **3**, 113–27.

Kutty, M.N. & Sukumaran, N. (1975). Influence of upper and lower temperature extremes on the swimming performance of *Tilapia mossambica*. *Transactions of the American Fisheries Society*, **104**, 755–61.

Laird, P.V., Baud, W.C. & Hoffman, R.B. (1972). Brain extracts from donors exposed to inescapable stressors modify shock-avoidance learning rate in recipient goldfish. *Journal of Biological Psychology*, **14**, 8–11.

Laudien, H., Freyre, J., Erb, R. & Denzer, D. (1986). Influence of isolation stress and inhibited protein biosynthesis on learning and memory in goldfish. *Physiology and Behaviour*, **38**, 621–8.

Lemly, A.D. & Smith, R.J.F. (1987). Effects of chronic exposure to acidified water on chemoreception of feeding stimuli in fathead minnow (*Pimephales promelas*): mechanisms and ecological implications. *Environmental Toxicology and Chemistry*, **6**, 225–38.

Leshner, A.I. (1978). *An introduction to behavioral endocrinology*. p. 361 Oxford University Press, New York.

Lewis, D.H., Tarpley, R.J., Marks, J.E. & Sis, R.F. (1985). Drug induced structural changes in olfactory organ of channel catfish *Ictalurus punctatus*, Rafinesque. *Journal of Fish Biology*, **26**, 355–8.

Li, H.W. & Brocksen, R.W. (1977). Approaches to the analysis of energetic costs of intraspecific competition for space by rainbow trout (*Salmo gairdneri*). *Journal of Fish Biology*, **11**, 329–41.

Løkkeborg, S., Olla, B.L., Pearson, W.H. & Davis, M.W. (1995). Behavioural responses of sablefish, *Anoplopoma fimbria*, to bait odour. *Journal of Fish Biology*, **46**, 142–55.

Magnuson, J.J. (1962). An analysis of aggressive behavior, growth, and competition for food and space in medaka *(Oryzias latipes) (Pisces, Cyprinodontidae)*. *Canadian Journal of Zoology*, **40**, 313–63.

Marcucella, H., Abramson, C.I. (1978). Behavioral toxicology and teleost fish. In *The Behavior of Fish and Other Aquatic Animals*. Mostofsky, D.I. (ed.) pp. 33–77. Academic Press, New York.

Maule, A.G., Schreck, C.B., Bradford, C.S. & Barton, B.B. (1988). Physiological effects of collecting and transporting emigrating juvenile chinook salmon past dams on the Columbia River. *Transactions of the American Fisheries Society*, **117**, 245–61.

Mazeaud, M.M., Mazeaud, F. & Donaldson, E.M. (1977). Primary and secondary effects of stress in fish: some new data with a general review. *Transactions of the American Fisheries Society*, **16**, 201–12.

McNicholl, P.G. & Mackay, W.C. (1975*a*). Effect of DDT and M.S. 222 on learning a simple conditioned response in rainbow trout *(Salmo gairdneri)*. *Journal of the Fisheries Research Board of Canada*, **32**, 661–5.

McNicholl, P.G. & Mackay, W.C. (1975*b*). Effect of DDT on discriminating ability of rainbow trout *(Salmo gairdneri)*. *Journal of the Fisheries Research Board of Canada*, **32**, 785–8.

Mesa, M.G. (1994). Effects of multiple acute stressors on the predator avoidance ability and physiology of juvenile chinook salmon. *Transactions of the American Fisheries Society*, **123**, 786–93.

Mesa, M.G. & Schreck, C.B. (1989). Electrofishing mark-recapture and depletion methodologies evoke behavioral and physiological changes in cutthroat trout. *Transactions of the American Fisheries Society*, **118**, 644–58.

Metcalfe, N.B. (1986). Intraspecific variation in competitive ability and food intake in salmonids: consequences for energy budgets and growth rates. *Journal of Fish Biology*, **20**, 525–31.

Metcalfe, N.B., Huntingford, F.A. & Thorpe, J.E. (1987). The influence of predation risk on the feeding motivation and foraging strategy of juvenile Atlantic salmon. *Animal Behaviour*, **35**, 901–11.

Nakamura, F. (1986). Avoidance behavior and swimming activity of fish to detect pH changes. *Bulletin of Environmental Contamination and Toxicology*, **37**, 808–15.

de-Oliveira-Fernandes, M. & Volpato, G.L. (1993). Heterogeneous growth in the Nile tilapia: social stress and carbohydrate metabolism. *Physiology and Behaviour*, **54**, 319–23.

Olla, B.L. & Davis, M.W. (1989). The role of learning and stress in predator avoidance of hatchery-reared coho salmon *(Oncorhynchus kisutch)* juveniles. *Aquaculture*, **76**, 209–14.

Olla, B.L. & Davis, M.W. (1992). Phototactic responses of unfed walleye pollock, *Theragra chalcogramma* larvae: comparisons with other measures of condition. *Environmental Biology of Fishes*, **35**, 105–8.

Olla, B.L. & Studholme, A.L. (1971). The effect of temperature on the activity of bluefish, *Pomatomus saltatrix* L. *Biological Bulletin*, **141**, 337–49.

Olla, B.L. & Studholme, A.L. (1975). The effect of temperature on the behavior of young tautog, *Tautoga onitis* (L.). *Proceedings of the Ninth European Marine Biology Symposium*. Barnes, H. (ed.) pp. 75–93. Aberdeen University Press, Aberdeen.

Olla, B.L., Studholme, A.L., Bejda, A.J., Samet, C. & Martin, A.D. (1975). The effects of temperature on the behavior of marine fishes: a comparison among Atlantic mackerel, *Scomber scombrus*, bluefish,

Pomatomus saltatrix, and tautog, *Tautoga onitis.* In *Combined Effects of Radioactive, Chemical and Thermal Releases to the Environment.* pp. 33–77. International Atomic Energy Agency, SM-197/4, Vienna.

Olla, B.L., Studholme, A.L., Bejda, A.J., Samet, C. & Martin, A.D. (1978). Effect of temperature on activity and social behavior of the adult tautog *Tautoga onitis* under laboratory conditions. *Marine Biology*, **45**, 369–78.

Olla, B.L., Pearson, W.H. & Studholme, A.L. (1980*a*). Applicability of behavioral measures in environmental stress assessment. *Rapports et Proces-Verbaux des Reunions Conseil International pour L'Exploration de la Mer*, **179**, 162–73.

Olla, B.L., Atema, J., Forward, R., Kittredge, J., Livingston, R.J., McLeese, D.W., Miller, D.C., Vernberg, W.B., Wells, P.G. & Wilson, K. (1980*b*) The role of behavior in marine pollution monitoring. *Rapports et Proces-Verbaux des Reunions Conseil International pour L'Exploration de la Mer*, **179**, 174–81.

Olla, B.L., Studholme, A.L. & Bejda, A.J. (1985). Behavior of juvenile bluefish *Pomatomus saltatrix* in vertical thermal gradients: influence of season, temperature acclimation and food. *Marine Ecology Progress Series*, **23**, 165–77.

Olla, B.L., Davis, M.W. & Schreck, C.B. (1992). Comparison of predator avoidance capabilities with corticosteroid levels induced by stress in juvenile coho salmon. *Transactions of the American Fisheries Society*, **121**, 544–7.

Olla, B.L., Davis, M.W. & Ryer, C.H. (1994). Behavioural deficits in hatchery reared fish: potential effects on survival following release. *Aquaculture Fisheries Management*, **25**, Supplement 1, 19–34.

Olla, B.L., Davis, M.W. & Schreck, C.B. (1995). Stress-induced impairment of predator evasion and non-predator mortality in Pacific salmon. *Aquaculture Research*, **26**, 393–8.

Olla, B.L., Davis, M.W., Ryer, C.H. & Sogard, S.M. (1996). Behavioral determinants of distribution and survival in early stages of walleye pollock, *Theragra chalcogramma*: a synthesis of experimental studies. *Fisheries Oceanography*, **5** Supplement 1, 167–78.

Parenskiy, V.A. (1989). The spawning success of sockeye salmon, *Oncorhynchus nerka*, as dependent on the behaviour pattern of spawners on spawning grounds. *Journal of Ichthyology*, **29**, 985–93.

Parenskiy, V.A. (1990). Relation between the spawning success of sockeye salmon, *Oncorhynchus nerka* and behavior on spawning grounds. *Journal of Ichthyology*, **30**, 48–58.

Pazkowski, C.A. & Olla, B.L. (1985). Social interactions of coho salmon (*Oncorhynchus kisutch*) smolts in seawater. *Canadian Journal of Zoology*, **63**, 2401–7.

Pearson, W.H., Miller, S.E. & Olla, B.L. (1980). Chemoreception in the food searching and feeding behavior of the red hake, *Urophycis chuss* (Walbaum). *Journal of Experimental Marine Biology and Ecology*, **48**, 139–50.

Pearson, W.H., Sugarman, P.C., Woodruff, D.L. & Olla, B.L. (1981). Impairment of the chemosensory antennular flicking response in the dungeness crab, *Cancer magister*, by petroleum hydrocarbons. *Fisheries Bulletin*, **79**, 641–7.

Peterson, R.H. (1976). Temperature selection of juvenile Atlantic salmon (*Salmo salar*) as influenced by various toxic substances. *Journal of the Fisheries Research Board of Canada*, **33**, 1722–30.

Pickering, A.D., Pottinger, T.G. & Christie, P. (1982). Recovery of the brown trout, *Salmo trutta* L., from acute handling stress: a time-course study. *Journal of Fish Biology*, **20**, 229–44.

Pottinger, T.G. & Pickering, A.D. (1992). The influence of social interaction on the acclimation of rainbow trout, *Oncorhynchus mykiss* (Walbaum) to chronic stress. *Journal of Fish Biology*, **41**, 435–47.

Quinn, T.P., Olson, A F. & Konecki, J.T. (1988). Effects of anesthesia on the chemosensory behaviour of Pacific salmon. *Journal of Fish Biology*, **33**, 637–41.

Rand, G.M. (1985). Behavior. In *Fundamentals of Aquatic Toxicology Methods and Applications* Rand, G.M. & Petrocelli, S.R. (eds.) pp. 221–63. Hemisphere Publishing Corporation, New York.

Rehnberg, B.G., Smith, R.J.F. & Sloley, B.D. (1987). The reaction of pearl dace (*Pisces, Cyprinidae*) to alarm substance: time-course of behavior, brain amines, and stress physiology. *Canadian Journal of Zoology*, **65**, 2916–21.

Reynolds, W.W. (1977). Fever and antipyresis in the bluegill sunfish, *Lepomis macrochirus*. *Comparative Biochemistry and Physiology*, **57C**, 165–7.

Reynolds, W.W., Covert, J.B. & Casterlin, M.E. (1978). Febrile responses of goldfish *Carassius auratus* L. to *Aeromonas hydrophila* and to *Escherichia coli* endotoxin. *Journal of Fish Diseases*, **1**, 217–73.

Ryer, C.H. & Olla, B.L. (1991). Agonistic behavior in a schooling fish: form, function and ontogeny. *Environmental Biology of Fishes*, **31**, 355–63.

Sapolsky, R.M., Krey, L.C. & McEwen, B.S. (1984a). Stress down-regulates corticosterone receptors in a site-specific manner in the brain. *Endocrinology*, **114**, 287–92.

Sapolsky, R.M., Krey, L.C. & McEwen, B.S. (1984b). Glucocorticoid-sensitive hippocampal neurons are involved in terminating the adrenocortical stress response. *Proceedings of the National Academy of Sciences of the USA*, **81**, 174–77.

Scherer, E. (1992). Behavioural responses as indicators of environmental alterations: approaches, results, developments. *Journal of Applied Ichthyology*, **8**, 122–31.

Schreck, C.B. (1981). Stress and compensation in teleostean fishes: response to social and physical factors. In *Stress and Fish*. Pickering, A.D. (ed.) pp. 295–321. Academic Press, London.

Schreck, C.B. (1982). Stress and rearing of salmonids. *Aquaculture*, **28**, 341–9.

Schreck, C.B. (1990). Physiological, behavioral, and performance indicators of stress. *American Fisheries Society Symposium*, **8**, 29–37.

Schreck, C.B. (1992). Glucocorticoids: metabolism, growth, and development. In *The Endocrinology of Growth, Development and Metabolism in Vertebrates*. Schreibman, M.P., Scanes, C.G. & Pang, P.K.T. (eds.) pp. 367–92. Academic Press, New York.

Schreck, C.B. & Li, H.W. (1991). Performance capacity of fish: stress and water quality. In *Advances in World Aquaculture*, Volume 3. *Aquaculture and Water Quality*. Brune, D.E. & Tomasso, J.R. (eds.) pp. 27–9. World Aquaculture Society, Baton Rouge, Louisiana.

Schreck, C.B., Jonsson, L., Feist, G. & Reno, P. (1995). Conditioning improves performance of juvenile Chinook salmon, *Oncorhynchus tshawytscha*, to transportation stress. *Aquaculture*, **135**, 99–100.

Schreck, C.B., Solazi, M.F., Johnson, S.L. & Nickelson, T.E. (1989). Transportation stress affects performance of coho salmon, *Oncorhynchus kisutch*. *Aquaculture*, **82**, 15–20.

Selye, H. (1936). A syndrome produced by diverse nocuous agents. *Nature*, **138**, 32.

Selye, H. (1950). Stress and the general adaptation syndrome. *British Medical Journal*, **1**, 1383–92.

Semenchenko, N.N. (1988). Mechanisms of innate population control in sockeye salmon, *Oncorhynchus nerka*. *Journal of Ichthyology*, **28**, 149–57.

Shashoua, V.E. (1973). Seasonal changes in learning activity patterns of goldfish. *Science*, **181**, 572–4.

Sigismondi, L.A. & Weber, L.J. (1988). Changes in avoidance response time of juvenile chinook salmon exposed to multiple acute handling stresses. *Transactions of the American Fisheries Society*, **117**, 196–201.

Sogard, S.M. & Olla, B.L. (1996). Food deprivation affects vertical distribution and activity of walleye pollock, *Theragra chalcogramma*, juveniles in a thermal gradient: potential energy conserving mechanisms. *Marine Ecology Progress Series*, **133**, 43–55.

Srivastava, R.K., Brown, J.A. & Allen, J. (1991). The influence of wave frequency and wave height on the behavior of rainbow trout (*Oncorhynchus mykiss*) in cages. *Aquaculture*, **97**, 143–53.

Sullivan, J.F., Atchison, G.J., Kolar, D.J. & McIntosh, A.W. (1978). Changes in the predator-prey behavior of fathead minnows (*Pimephales promelas*) and largemouth bass (*Micropterus salmoides*) caused by cadmium. *Journal of the Fisheries Research Board of Canada*, **35**, 446–51.

Taylor, P.B. (1988). Effects of anaesthetic MS 222 on the orientation of juvenile salmon, *Oncorhynchus tschawytscha* Walbaum. *Journal of Fish Biology*, **32**, 161–8.

Thoesen, J.C. (ed.) (1994). *Suggested procedures for the detection and identification of certain fin fish and shellfish pathogens.* 4th edition, Version 1. Fish Health Section, American Fisheries Society, Bethesda, Maryland.

Tsai, C.-L. & Hoh, K-H. (1995). Effect of indomethacin on survival of *Aeromonas hydrophila* infected tilapia, *Oreochromis mossambicus*. *Zoological Studies*, **34**, 59–61.

Vosyliene, M.-Z., Petrauskiene, L. & Prekeris, R. (1993). Behavioural responses and physiological parameters of trout at various stages of social stress. *Biologija*, **2**, 86–90.

Wardle, C.S. (1986). Fish behaviour and fishing gear. In *The Behaviour of Teleost Fishes*. Pitcher, T.J. (ed.) pp. 463–95. Croom Helm, London.

Warren, J.W. (1991). *Diseases of Hatchery Fish*. p. 92. United States Department of Interior, Fish and Wildlife Service, Washington, DC.

Woltering, D.M., Hedtke, J.L. & Weber, L.J. (1978). Predator–prey interactions of fishes under the influence of ammonia. *Transactions of the American Fisheries Society*, **107**, 500–4.

T.G. POTTINGER and A.D. PICKERING

Genetic basis to the stress response: selective breeding for stress-tolerant fish

Introduction

Stress is an unavoidable component of the finfish aquaculture environment. This is primarily because ideal husbandry practices are, of necessity, compromised by the economic constraints of large-scale fish production. The aquaculture environment exposes fish to a regime of repeated acute stress (e.g. handling, grading, transport, prophylactic treatment) and, in some instances, chronic stress (e.g. deterioration in water quality, overcrowding). In addition, the aquaculture environment is inherently unsuitable for fish that are territorial or solitary animals in their natural environment, such as some salmonid fish. In these cases, agonistic interactions can be particularly stressful to the fish. As described elsewhere in this volume (see chapter by B. Barton) the physiological response of fish to a stressful stimulus is essentially adaptive and results in modifications to cardiorespiratory and metabolic functions which enhance the fish's likelihood of survival under challenging conditions. It is well established that when exposed to repeated, acute or to chronic challenge, the stress response of fish ceases to be adaptive and becomes, instead, maladaptive. In such circumstances substantial deleterious effects on growth (Pickering, 1993), reproductive function (Campbell, Pottinger & Sumpter, 1992, 1994) and immunocompetence (Pickering & Pottinger, 1989) are observed. In addition, stress prior to slaughter may have adverse effects on flesh quality, in particular with respect to muscle adenosine triphosphate (ATP) levels, the rate of onset of rigor mortis, and 'gaping' of the tissue (Lavety, 1993; Lowe *et al.*, 1993). Domestication, which can be considered to be *'that process by which a population of animals become adapted to humans and to the captive environment by some combination of genetic changes occurring over generations and environmentally induced developmental events re-occurring during each generation'* (Price, 1984), has, in most cases, not advanced as far with cultured fish as with other species of animal exploited by agriculture.

The purpose of this chapter is to consider the possibility that the responsiveness of fish to stress is a feature which has a distinct genetic component and may therefore be modified by selective breeding. By increasing the tolerance of fish to stress, thereby ameliorating some of the effects of unavoidable stress, it may be feasible to generate strains of fish that display improved performance within the aquaculture environment, across a number of traits.

The basic assumption underlying this contention is that, within an environment in which fish are exposed to frequent or prolonged episodes of stress, those individuals with a low level of responsiveness to stress will be at an advantage relative to fish displaying a higher level of responsiveness to stress. Such differences might be envisaged as arising through a divergence between individuals in the *threshold* at which stimuli evoke a stress response, in the *magnitude* or *duration* of the response once initiated, or even in a reduction in the *sensitivity* of target tissues to the physiological perturbations initiated by stress, or a combination of all these. Given the broadly adverse effects of repeated acute, or chronic, stress, an enhanced tolerance of stressful procedures might be postulated to result in improved food conversion efficiency and growth, and in a reduction in the incidence of disease. An improvement in the performance of broodstock, in terms of fecundity, egg quality and post-spawning survival, is also likely, together with an improvement in post-slaughter flesh quality. All of these traits may be considered by the aquaculture industry as key candidates for improvement. In addition, and perhaps of equal importance, an attendant benefit of reducing the adverse effects of stress under aquaculture conditions is an improvement in the 'well-being' of farmed fish. This would be achieved primarily by reducing the incidence of disease and thus alleviating disease-related suffering, but it is possible that some of the adverse effects of unavoidable social interaction (aggression, dominance) would also be attenuated. A reduction in the requirement for prophylactic and therapeutic agents would also have financial and environmental benefits.

The working hypothesis that reducing stress responsiveness results in an overall improvement in performance assumes that the beneficial aspects of the stress response are outweighed by the damaging effects of stress under aquaculture conditions. In fish destined for release into the natural environment, the converse may be true. In addition, a straightforward inverse relationship between high stress responsiveness and adverse effects is assumed in the first instance, although it is recognized that this may well be an overly simplistic analysis of the situation, particularly with regard to the impact of stress on the immune system (Mason, 1991).

The application of selective breeding in aquaculture

The current status of selective breeding in aquaculture

Selective breeding is a strategy which has been employed for centuries, both deliberately and inadvertently, to enhance and emphasize traits seen as desirable in animals. However, although selective breeding of cyprinids for morphological features has long been practised in Asia, it is only relatively recently that selection for traits considered to be important in fish reared for food, has been undertaken. The genetic principles underlying selective breeding in fish are described thoroughly by Tave (1993) and Purdom (1993).

It is outside the scope of this chapter to examine in detail or at length the whole field of selective breeding, or genetic improvement, in fish. Species in which genetic variation of desirable traits has been specifically studied are predominantly those of aquacultural significance; rainbow trout (*Oncorhynchus mykiss*; Gjedrem, 1992), Pacific salmon (various *Oncorhynchus spp.*; Withler & Beacham, 1994), Atlantic salmon (*Salmo salar*; Rye, Lillevik & Gjerde, 1990), channel catfish (*Ictalurus punctatus*; Dunham, Smitherman & Goodman, 1987), and tilapia (*Oreochromis niloticus*; Eknath *et al.*, 1993). Particular traits that have been targeted for improvement or genetic analysis include body weight/growth (Gall & Huang, 1988*a*), food conversion efficiency (Kinghorn, 1983), female reproductive performance (Gall & Huang, 1988*b*), male reproductive performance (Blanc *et al.*, 1993), age at first maturation (Burger & Chevassus, 1987), cold tolerance (Tave, Smitherman & Jayaprakas, 1989), and disease resistance (Chevassus & Dorson, 1990). In some cases, the chosen traits have proved refractory to improvement, or unsuitable as the basis of a selection programme, as was noted for food conversion efficiency in rainbow trout (Kinghorn, 1983). However, some degree of success has been achieved in enhancing other traits in this species, such as body weight at a given age (Hörstgen-Schwark, 1993).

The genetic improvement of disease resistance in fish has received a considerable amount of attention, because of the major impact of disease outbreaks on farmed fish stocks. The prospects for, and problems associated with, increasing disease resistance are reviewed by Fjalestad, Gjedrem and Gjerde (1993). Several studies have demonstrated the existence of genetic variation in resistance to diseases including furunculosis (Gjedrem, Salte & Gjöen, 1991; Gjedrem & Gjöen, 1995), cold water vibriosis and bacterial kidney disease (BKD; Gjedrem & Gjöen, 1995) and fungal infection (*Saprolegnia* sp.; Nilsson, 1992) in salmonids, although others have failed to find such a relationship (Beacham & Evelyn, 1992).

It is in the context of selection for improvements in disease resistance that the stress response in fish was first considered as a trait upon which selection pressure should be directed. Because mortality rate has been suggested to be an inadequate indicator of disease susceptibility (Fevolden, Refstie & Røed, 1991), and because of the well-established links between the stress response and immunosuppression, the use of the stress response as an indirect selection criterion for disease resistance has been proposed (Refstie, 1982, 1986; Fevolden *et al.*, 1991).

Methodologies employed in selective breeding of fish

Selection is applied to a breeding population in an attempt to manipulate the population mean for the trait of interest. Tave (1993) outlines the theory underlying selective breeding programmes. The variance for any quantitative phenotype (V_P) is the sum of the genetic variance (V_G), the environmental variance (V_E), and the interaction between genetic and environmental variance (V_{G-E}):

$$V_P = V_G + V_E + V_{G-E}$$

It is the genetic variance that is the object of interest in selective breeding. Genetic variance can be subdivided into several component parts (additive genetic variance V_A, dominance genetic variance V_D, epistatic genetic variance V_I) but, of these, it is the *additive genetic variance* (V_A) that is exploited during selection (see Tave, 1993, for further details). Additive genetic variance represents the total effect of all alleles across all loci, independent of specific interactions or combinations of alleles. Heritability (h^2) is the proportion of phenotypic variance that is dependent on genetic variance and is inherited in a predictable fashion (Tave, 1993) and it is in terms of heritability estimates that the results of most selective breeding studies are reported (Gjedrem, 1992).

Gall (1990) and Refstie (1990) outline the basic requirements of a selective breeding programme and describe appropriate methodologies. In most cases fish populations are subjected to directional selection, in which the aim is to improve a specific trait. Selection of individuals from which progeny will be generated may be achieved in several ways (Gall, 1990).

- *Pedigree selection* employs data on the performance of ancestors.
- *Progeny testing* assesses the performance of individuals on the basis of their offspring.

- *Family selection* employs assessment and comparison of the performance of whole families, rather than of individuals.
- *Sib selection* utilizes the performance of groups of siblings to estimate the breeding performance of a family member.
- *Combined selection* employs more than one method of estimating breeding performance, for example, combining between-family and within-family selection.

Combined selection is considered to be the most effective method of achieving the aims of selective breeding programmes (Gall & Huang, 1988*a*, *b*; Reftsie, 1990).

There can be drawbacks when attempting the selective improvement of a single trait (Tavc, 1993). Selection for a single phenotype may inadvertently co-select for an undesirable phenotype. For example, it was found that selection for rapid growth was unavoidably linked with early maturation in Atlantic salmon (Thorpe *et al.*, 1983). Breeding for improvement in more than one trait (tandem selection) is inefficient because of the time required to select for a single phenotype over several generations, followed by the time required to select for additional phenotypes over subsequent generations. The alternative to tandem selection, independent culling, can result in the elimination of valuable alleles and a reduction in population size. It is clearly desirable to identify a single trait, the improvement of which will have the widest possible benefits in terms of improved performance. This is the rationale underlying current efforts to selectively breed fish for low stress responsiveness. The likelihood that this approach will be successful can be assessed by consideration of work carried out using poultry; these animals are reared under an intensive farming system which has much in common with intensive finfish aquaculture.

Manipulation of stress responsiveness by selective breeding

Modification of the stress response by selective breeding of poultry

It is appropriate to summarize briefly the extent of work in this area which has been directed towards reducing the stress responsiveness of poultry as, by analogy, these data lend credence to the assertion that the magnitude of the stress response in fish may have a genetic basis. The potential impact of stress on poultry in intensive rearing environments, and the possible advantages of a low response to stress were being discussed almost 40 years ago (Brown, 1959). The results of early attempts to breed selectively lines of turkey for divergent responsiveness to stress are reviewed by Brown and Nestor (1974). Cold stress was utilized as the selection tool and mean plasma corticosterone levels were used to measure the magnitude of the stress response. The selection procedure was continued for nine generations, resulting in significant divergence of the plasma corticosterone response to stress between the two lines (indeed, divergence was apparent within one generation). These studies highlight one advantage of working with poultry rather than fish, that is, the relatively short generation times of birds compared to those of most fish (weeks cf. years). High-responding birds were found to be more excitable than low responders and exhibited slower growth. 'Natural mortality' was found to be lower in the low-responding birds than in the high-responding line but the effect on disease susceptibility was equivocal, with the authors citing evidence to show that although high-responding birds are more susceptible to some pathogens (particularly viral infections) this is not a universal effect for all disease agents. Subsequent work on the Japanese quail (*Coturnix coturnix japonica*), which has an even shorter generation time, resulted in greater divergence in the corticosterone response to stress of lines selected via exposure to immobilization stress (Satterlee & Johnson, 1988) over 12 generations. Most recently, research in this area has focused on 'fear' and 'distress' in the quail in relation to pituitary–adrenal activation, the rationale being that reduced fearfulness of certain stimuli is an advantage in the intensive rearing environment (Jones, Satterlee & Ryder, 1992, 1994b; Jones et al., 1994a). If behavioural correlates of stress responsiveness can be identified, then they may provide a more convenient and, perhaps, more reliable index of stress responsiveness than physiological measurements.

Clearly, the results obtained with poultry strongly suggest that attempts to breed selectively for an attenuated responsiveness to stress in fish are likely to be successful. The greatest drawback in attempting to achieve this goal, however, is the generation time associated with those fish species of aquacultural significance.

Modification of the stress response of fish by selective breeding

It is clear that an essential prerequisite for undertaking a programme of breeding designed to improve a specific trait is to demonstrate or establish that a significant amount of variance within the trait is genetic in origin and does not arise wholly as a result of environmental factors.

There is some evidence that the responsiveness of fish to stress is, at least in part, genetically influenced.

The measure which has been most widely employed as a marker of responsiveness to stress in fish is the extent to which plasma cortisol levels are elevated following a stressful stimulus. Cortisol is a reliable indicator of the activation of the hypothalamic–pituitary–interrenal axis (see chapter by B. Barton, this volume) and, given that cortisol is implicated in many of the adverse effects of chronic stress (see chapters by B. Barton, G. Van Der Kraak, P.H.M. Balm, this volume), the extent and duration of plasma cortisol elevation following stress can be considered to be related to the deleterious effect associated with the stressful stimulus. Although this assumption is somewhat simplistic, given that it does not take into account variation in tissue sensitivity to cortisol and other endocrine influences, it has been employed as the basis for the majority of published comparisons of stress responsiveness in fish. However, it must be emphasized that the extent to which plasma cortisol levels are elevated during stress is subject to extrinsic and intrinsic influence. Water temperature (Sumpter, Pickering & Pottinger, 1985), water quality (Pickering & Pottinger, 1987) and sexual maturity (Pottinger, Balm & Pickering, 1995) all modulate the corticosteroid response to stress in salmonid fish. Despite these additional sources of variation, there are data suggesting that fish of the same species, but of different genetic origin, can differ markedly in their corticosteroid responsiveness to stress.

Strain differences in the response of fish to stress

Consistent differences in the *magnitude* of the cortisol response were observed among five strains of hatchery-reared rainbow trout maintained under identical conditions and exposed to a standardized confinement stress over an 8-month period (Pickering & Pottinger, 1989). Marked differences in the *dynamics* of the plasma cortisol and cortisone responses to confinement stress have also been noted to occur between strains of rainbow trout (Pottinger & Moran, 1993). In a study in which cortisol levels were not measured, but a variety of physiological indices were quantified following stressful challenge tests, significant differences in the response to a challenge were noted to occur among six genetically distinct strains of coho salmon (McGeer, Baranyi & Iwama, 1991).

There are also a number of reports of differences in the response to stress between wild and hatchery-reared salmonids (Woodward & Strange, 1987; Salonius & Iwama, 1991). However, the difficulty in interpreting these data lies in distinguishing between genetic and environmental influences on stress responsiveness. It is clear that environmental effects, for example the conditions under which the fish are reared, can have a substantial impact on the responsiveness of fish to stress. Salonius and Iwama (1993) found that coho (*Onchorhynchus kisutch*) and chinook salmon (*O. tshawytscha*), hatchery reared from eggs collected from naturally spawning wild fish, showed significantly lower cortisol responses to stress than their free-living counterparts despite having been derived from the same genetic stock (Fig. 1). Furthermore, coho salmon collected as eggs from the natural environment, but reared under hatchery conditions for 5 months and then transferred to the natural riverine environment as fry, showed a cortisol response to stress similar to that of wild fish (Fig. 1), i.e higher than that shown by their hatchery-maintained siblings. This difference in response level between hatchery-maintained and transplanted fish of the same stock was maintained even after the transplanted and wild fish were reared under hatchery conditions for 7 months. The authors speculated that frequent exposure to stress in the hatchery environment at a key developmental stage may be responsible for modifying the sensitivity of hatchery-reared fish to stress or, more simply, that frequent exposure to stress 'down-regulates' responsiveness. The fact that wild fish retain their level of responsiveness to stress even after 7 months of captivity supports the assertion that experience or environment only modify responsiveness during a particular developmental 'window'. There is considerable evidence that, in mammals, exposure to stress during early development can permanently affect subsequent responsiveness to stressful stimuli (Meaney *et al.*, 1993). However, if early experience is crucial, then it is surprising that fish initially reared in

Fig. 1. Plasma cortisol levels in juvenile coho and chinook salmon (mean ± SEM, $n = 12$) at 0, 4, 8 and 30 h following a 30- to 60-s dipnet handling stress. The fish all originated from the same group of spawning adults. Wild fish developed from eggs spawned naturally in the river. Colonized fish were artificially propagated from eggs and milt removed from spawning adults and were maintained in a hatchery environment for 5 months post-hatch before release into the upper watershed of the river as fry. Hatchery fish were obtained from the same source as colonized fish but were maintained under hatchery conditions for the entire rearing period. Different superscripts denote significant differences ($p \leq 0.05$) among means. (Figure adapted from Salonius & Iwama, 1993. Used with permission.)

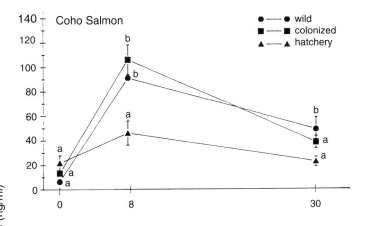

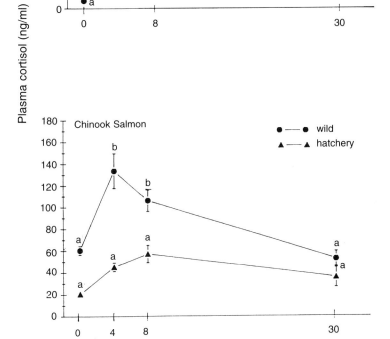

Time post-stress (h)

the hatchery environment for 5 months, released into the natural
environment for an indeterminate period and then transferred to the
hatchery environment for a further 7 months showed a response level
comparable to that of wild fish (Salonius & Iwama, 1993).

Individual differences in the response of fish to stress

If a combined selection method is deemed the most effective approach
to selective breeding for reduced stress responsiveness, it is necessary
to assess the response of individuals, as well as of groups of fish, to
a stressful stimulus. It has been reported that variation in post-stress
cortisol levels between individuals within a single family is low by
comparison with the variation observed in randomly chosen groups of
fish (Heath et al., 1993). While a single 'stress test' applied to a large
sample of fish may be adequate to determine the responsiveness of a
population, is the response of individual fish consistent enough with
time to permit the same approach when attempting to identify fish
displaying the desired level of responsiveness? Few of the studies
described so far have examined the responsiveness to stress of the
populations under examination on more than one occasion and none
have made repeated measures of individual fish. In order to establish
whether stress responsiveness is a characteristic that is stable with time,
and to assess the within-population fish-to-fish variation in responsiveness, Pottinger, Pickering and Hurley (1992) subjected a large group
of rainbow trout from each of two strains to a standardized confinement
stress at intervals over a 6-month period. The post-stress cortisol levels
of each individual fish were determined on a total of nine occasions
over a 12-month period. The authors found that only a limited number
of individuals (30%) displayed a consistent response to stress. These
fish were designated as either 'high responders' (HR) or 'low
responders' (LR), depending on whether the magnitude of their
response was consistently within the upper or lower 50% range of the
population as a whole (Fig. 2). A considerable degree of divergence
in the magnitude of mean post-stress cortisol levels of HR and LR
fish was observed over several years, with post-stress cortisol levels in
the HR group being, on average, twice those of the LR group. The
authors concluded that the level of response of the selected fish was
not environmentally determined as response levels remained consistent,
for up to 28 months, even when the rearing conditions were altered.
Similar results were obtained for the stability of the catecholamine
response to stress in rats, in which the relative magnitude of the
individual stress response was found to be stable over a 12-month
period (Taylor et al., 1989).

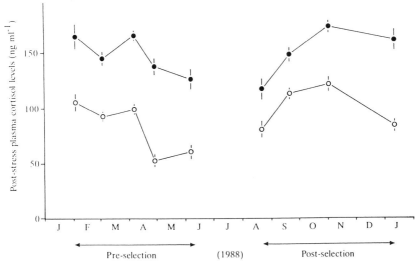

Fig. 2. Plasma cortisol levels after 1 h of confinement stress in rainbow trout selected for high (●) and low (○) responsiveness to stress. Pre-selection refers to the period prior to their removal from the stock population, during which selection occurred. Post-selection indicates the period during which fish of similar response levels were segregated from the stock population and maintained together. Each point is the mean ± SEM, n = 34–36. (Figure adapted from Pottinger, Pickering & Hurley, 1992. Used with kind permission of Elsevier Science Publishers.)

Evidence that stress responsiveness may be genetically determined in fish

An empirical approach was employed to assess whether rainbow trout identified as HR or LR to stress in an earlier study (Pottinger *et al.*, 1992) would generate progeny with similar tendencies. The fish selected in the previous study were employed to carry out a series of pooled gamete matings, HR females × HR males and LR females × LR males (Pottinger, Moran & Morgan, 1994). The progeny of these crosses were subjected to a standardized 1-h confinement stress at monthly intervals, for a period of 1 year. Overall, the progeny of high-responding parents displayed a significantly greater cortisol elevation under these conditions compared to the progeny of low-responding parents. When the time course of the cortisol response to continuous confinement was examined, the most pronounced differences in cortisol levels were observed 4 hours after the onset of confinement (Fig. 3). Stress-induced

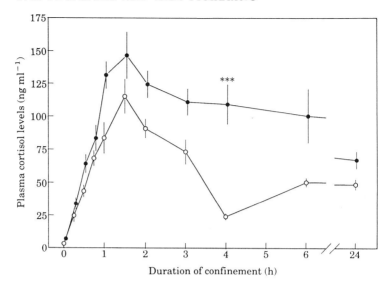

Fig. 3. Plasma cortisol levels at various time points following the onset of confinement stress in the progeny of high-responding (●, HR) and low-responding (○, LR) rainbow trout. Each point is the mean ± SEM, $n = 12$. Analysis of variance indicates that overall there is a significant difference ($p < 0.001$) between the cortisol response of HR and LR fish to confinement (HR = 89 ± 5 ng ml^{-1}; LR = 57 ± 4 ng ml^{-1}) although significant differences between specific time points are apparent only after 4 h (***, $p < 0.001$). (Figure adapted from Pottinger, Moran & Morgan, 1994. Used with permission.)

lymphocytopenia, a reduction in the number of circulating lymphocytes, is a cortisol-dependent phenomenon (Pickering, 1984) and examination of the effect of confinement stress on circulating lymphocyte numbers showed that HR progeny displayed a more sustained reduction in lymphocytes compared to LR fish, suggesting that differences in post-stress cortisol levels are of functional significance. Furthermore, the post-stress increase in circulating neutrophils was significantly greater in LR than HR fish. Of particular interest was the fact that highly significant differences in the response of circulating somatolactin (SL) levels were observed in the HR and LR progeny. HR fish showed a significantly greater elevation of SL in response to stress than did the LR group (Rand-Weaver, Pottinger & Sumpter, 1993). There is no evidence that either post-stress neutrophilia, or the SL response to stress is dependent upon the activity of the hypothalamic–pituitary–interrenal axis, suggesting that the selection process employed by these authors operated more broadly than originally anticipated. Although these results provide contributory evidence that stress responsiveness is transmitted across generations, this study suffers from inadequate information on the genetic pedigree of the fish employed.

An experimental design more appropriate for providing *quantitative* information on genetic influences on stress-related traits was employed by Heath *et al.* (1993). These authors undertook a detailed study which sought to clarify the interaction of environmental and genetic factors on apparent stress responsiveness in chinook salmon. Eight full-sib and half-sib families were maintained at two temperatures during egg development and, as fry, were subjected to handling stress. A number of somatic parameters were measured together with pre- and post-stress plasma cortisol and glucose levels. This study found that the cortisol response of these fish to an acute handling stress had a demonstrable genetic component; a significant sire effect. A significant maternal effect might also have been expected, but the small number of females employed in the study is suggested by the authors to have limited their ability to resolve dam effects. Nonetheless, both sire and dam effects were found to be significant factors influencing post-stress glucose levels. Some of the analytical procedures leading to these conclusions were subsequently queried (Silverstein, 1994). However, although statements regarding genetic–environment interactions were modified after re-analysis by the authors, such that only two stress-related traits (post-stress cortisol and glucose concentrations) were influenced by significant dam-by-environment interactions (Heath *et al.*, 1994), their assertion that the magnitude of the response to stress of progeny has, at the very least, a paternal component was still supported by the data. They concluded that the genetic and environmental background of fish can have significant effects on the stress response.

The most sustained effort directed at assessing the feasibility of selecting for stress responsiveness in fish has been that of Fevolden and colleagues. The underlying rationale for their work has been to utilize the stress response as an indirect marker of disease resistance. Mortality due to specific pathogens can show significant genetic variation (Fjalestad *et al.*, 1993), but the generation of strains which are resistant to more than one specific disease requires that selection criteria other than survival following a single pathogen challenge are employed (Fevolden *et al.*, 1992). Because of the well-established links between the stress response and the immune system, the response of fish to stress may provide an indirect measure of disease resistance. The use of an indirect selection method also offers the advantage of being measurable in live fish. This approach requires that a genetic link between responsiveness to stress and disease resistance be demonstrated (Fevolden *et al.*, 1991). These authors first sought to determine whether plasma cortisol levels are a suitable measure of a genetically differentiated stress response. They utilized a family selection scheme in which post-stress plasma cortisol and glucose levels were determined for more than 100 families of both rainbow trout and Atlantic salmon. Males from groups with high post-stress cortisol levels were crossed with females from other groups with high cortisol levels. Similar crosses were carried out for low responders. No sibs or half-sibs were crossed to avoid the effects of inbreeding. The progeny (F_1) of these crosses were themselves tested for stress responsiveness in their second year. The authors found that average values of post-stress plasma cortisol and glucose in Atlantic salmon were significantly different between the two lines, with higher levels of both variables in the high-response line than in the low-response line. For rainbow trout, only post-stress cortisol levels were significantly higher overall in the high-response line. However, both species showed highly significant sire effects on both plasma cortisol and plasma glucose levels. The authors concluded that their data gave support to the contention that the sensitivity of fish to environmental stress may have a genetic basis. The relatively small differences in the level of response between the two lines may have arisen because the selection procedure utilized to generate the high- and low-response lines in these experiments was not as efficient as it might have been. Because family (not combined) selection was employed it was not possible to take account of variation between individuals within families, in terms of the magnitude of the cortisol response or with regard to the consistency of the response of individuals with time.

The only quantitative estimates of h^2 for the cortisol and glucose responses to stress which are available to date were determined using the rainbow trout and Atlantic salmon populations that were employed to establish the high- and low-response lines already described (Fevolden *et al.*, 1991). Each year, for 4 years, fish from a number of full-sib families and half-sib groups were tested for stress responsiveness (Fevolden, Refstie & Gjerde, 1993*a*); 553 full-sib groups for the Atlantic salmon and 281 full-sib groups for the rainbow trout. Estimates of heritability across the year classes for both cortisol ($h^2 = 0.05$) and glucose ($h^2 = 0.03$) were not significantly different from zero in the salmon whereas, in the rainbow trout, heritability estimates for cortisol were of low to medium magnitude ($h^2 = 0.27$ across year classes) while heritability for glucose was low ($h^2 = 0.07$). It must be borne in mind when considering the apparently low heritability of these stress-related parameters that they do not take into account the evidence cited above (Pottinger *et al.*, 1992), i.e. that the magnitude of post-stress cortisol elevation may be a consistent and stable trait in only a proportion of the population. Fevolden *et al.* (1993*b*) suggest that more accurate estimates of the heritability of the stress response may be obtained by optimizing the design of the breeding programme, by employing a more standardized stressor and by using individuals with consistent response levels to stress to generate progeny.

The use of indices of stress other than plasma cortisol as a basis for selection

It has recently been suggested that post-stress plasma lysozyme levels may be a more reliable indicator of the physiological 'stress status' of salmonid fish compared to plasma cortisol levels (Fevolden & Røed, 1993). Levels of lysozyme, a bactericidal enzyme which hydrolyses bacterial cell wall mucopeptide linkages (Yousif, Albright & Evelyn, 1994), may increase or decrease following the onset of stress, depending on the nature of the stressor (Möck & Peters, 1990; Røed *et al.*, 1993). Post-stress lysozyme levels were found to be consistently higher in the high-response rainbow trout line than in the low-response line in one study in which post-stress cortisol levels failed to show a consistent relationship with selection line (Fevolden & Røed, 1993). In a further study designed to compare the genetic determinants of post-stress plasma cortisol and lysozyme in Atlantic salmon, fish were exposed to confinement stress for a minimum period of 30 min, and plasma cortisol and lysozyme levels were measured (Fevolden, Røed & Gjerde, 1994). This stressor caused an elevation of lysozyme levels for at least 1.5 h. It was found that the estimate of heritability for cortisol was low ($h^2 = 0.07$) and not significantly different from zero, whereas the heritability estimate for the post-stress lysozyme level was of medium magnitude ($h^2 = 0.19$). Furthermore, a negative correlation was observed between lysozyme activity and survival after challenge with *Aeromonas salmonicida* and *Renibacterium salmoninarum*, although not after challenge with *Vibrio salmonicida*. The authors suggest that high post-stress lysozyme activity is indicative of fish which possess a high level of stress responsiveness, and thus an increased susceptibility to disease, and that selection for a low post-stress lysozyme level is a suitable indirect criterion for selection for disease resistance.

The functional impact of selection for stress responsiveness in fish

The assumption underlying efforts to select fish for low responsiveness to stress is that such fish will show an enhanced performance under conditions of aquaculture. As yet, there is little evidence available with which to test this assumption. Pottinger *et al.* (1994) observed no difference in growth between progeny of the the high- and low-responding F_0 generation. However, these fish were maintained under conditions which were deliberately designed to minimize incidental stress – any advantages of possessing an attenuated stress response may only be apparent under more challenging conditions. To establish whether there is a genetic link between stress responsiveness and disease resistance, fish from the second (F_2) generation of the two lines of rainbow trout selected for high and low stress responsiveness (Fevolden *et al.*, 1991) were subsequently challenged with two different disease-causing organisms: *A. salmonicida*, the causative agent of furunculosis, and *V. anguillarum*, the causative agent of vibriosis (Fevolden *et al.*, 1992). Mortality rates were significantly greater among the high-response line exposed to *A. salmonicida* than among the low-response line. However, the reverse was true for the fish exposed to *V. anguillarum*. The authors concluded that selection for stress responsiveness also influenced disease resistance. They pointed out, however, that the fish used in the study were derived from a family selection scheme and that additional, unknown, traits may have been selected for.

A similar challenge experiment was carried out on the second generation of Atlantic salmon lines selected for low and high responsiveness to stress (Fevolden *et al.*, 1993*b*). In this case the fish were challenged with three bacterial pathogens: *A. salmonicida*, *V. salmonicida* and *R. salmoninarum*, the causative agent of BKD. The authors found that in both the furunculosis and vibriosis challenge tests significantly higher mortality was observed among fish of the high-stress response line (Fig. 4). No differences between lines were observed following the challenge with pathogens causing BKD. The results were confounded to an extent by the unexpected presence of *A. salmonicida* in fish challenged with pathogens causing vibriosis or BKD, which may have affected the response of the fish to the experimental challenge. However, as the authors noted, simultaneous exposure of fish to more than one pathogen is likely under aquaculture conditions, and the results are still strongly suggestive that specific disease resistance was affected by the selection scheme.

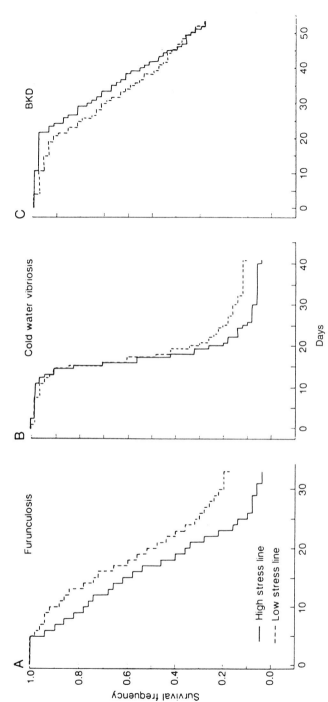

Fig. 4. The survival frequency of two lines of Atlantic salmon selected for a high or low stress response (based on the plasma cortisol response to confinement stress) and challenged with *Aeromonas salmonicida* (furunculosis), *Vibrio salmonicida* (cold water vibriosis), or *Renibacterium salmoninarum* (bacterial kidney disease, BKD). All challenges were carried out by intraperitoneal injection of the infective organism. There was a significant difference between the high and low stress lines, in terms of cumulative mortality, for the fish exposed to *A. salmonicida* and *V. salmonicida* ($p < 0.001$) but not for the fish exposed to *R. salmoninarum*. (Figure adapted from, Fevolden *et al.*, 1993b, with kind permission of Elsevier Science-Publishers.)

Overall, these results do not *conclusively* demonstrate a direct link between stress responsiveness and non-specific disease resistance. However, selection for stress responsiveness is clearly influencing susceptibility to specific diseases, suggesting that the sensitivity of fish to stress may ultimately prove to be a satisfactory indirect selection trait for disease resistance. As yet, strains of fish with stress responses which diverge in magnitude to an extent comparable to that achieved in poultry are unavailable. Assessment of the advantages (or disadvantages) of possessing a low response to stress will probably not be possible until such fish are available.

Conclusions

1. From the limited amount of work which has been carried out using fish, taken together with what is known of the genetic basis of stress responsiveness in mammals and birds, it seems likely that selective breeding for attenuated responsiveness to stress is feasible for fish.

2. The approach most likely to yield satisfactory results is a combined selection scheme, utilizing between- and within-family selection to identify breeding stock.

3. The choice of selection trait is crucial. Some doubts have been expressed regarding the efficacy of post-stress plasma cortisol as a selection trait on the basis of its apparently low heritability. However, current heritability estimates are open to criticism on the grounds that the methods used for assessing responsiveness to stress were imprecise. Further work is needed before cortisol can be dismissed as a potential index of responsiveness, particularly taking into account the great divergence achieved in post-stress corticosterone levels in poultry.

4. If the links between stress responsiveness and behaviour, which apparently exist in birds, can be demonstrated to occur in fish then it is possible that a non-invasive method for assessing stress-responsiveness in fish may be developed.

5. It has yet to be categorically demonstrated that possession of a high responsiveness to stress is a disadvantage and that low stress responsiveness is an advantage to fish under aquaculture conditions. The assumption that this is so is based on inference. Because individuals with divergent responsiveness to stress can be identified within populations by repeated testing, it might be appropriate to investigate further the effect of differing levels of responsiveness on a variety of parameters under aquaculture conditions.

References

Beacham, T.D. & Evelyn, T.P.T. (1992). Genetic variation in disease resistance and growth of chinook, coho, and chum salmon with respect to vibriosis, furunculosis, and bacterial kidney disease. *Transactions of the American Fisheries Society*, **121**, 456–85.

Blanc, J.M., Poisson, H., Escaffre, A.M., Aguirre, P. & Vallee, F. (1993). Inheritance of fertilizing ability in male tetraploid rainbow trout (*Oncorhynchus mykiss*). *Aquaculture*, **110**, 67–70.

Brown, K.I. (1959). 'Stress' and its implications in poultry production. *Worlds Poultry Science Journal*, **15**, 255–63.

Brown, K.I. & Nestor, K.E. (1974). Implications of selection for high and low adrenal response to stress. *Poultry Science*, **53**, 1297–306.

Burger, G. & Chevassus, B. (1987). Potential of selection for age at first sexual maturation in rainbow trout (*Salmo gairdneri*). *Bulletin Francais de la Peche et de la Pisciculture*, **307**, 102–17.

Campbell, P.M., Pottinger, T.G. & Sumpter, J.P. (1992). Stress reduces the quality of gametes produced by rainbow trout. *Biology of Reproduction*, **47**, 1140–50.

Campbell, P.M., Pottinger, T.G. & Sumpter, J.P. (1994). Preliminary evidence that chronic confinement stress reduces the quality of gametes produced by brown and rainbow trout. *Aquaculture*, **120**, 151–69.

Chevassus, B. & Dorson, M. (1990). Genetics of resistance to disease in fishes. *Aquaculture*, **85**, 83–107.

Dunham, R.A., Smitherman, R.O. & Goodman, R.K. (1987). Comparison of mass selection, crossbreeding, and hybridization for improving growth of channel catfish. *Progressive Fish Culturist*, **49**, 293–6.

Eknath, A.E., Tayamen, M.M., Palada-de Vera, M.S., Danting, J.C., Reyes, R.A., Dionisio, E.E., Capili, J.B., Bolivar, H.L., Abella, T.A., Circa, A.V., Bentsen, H.B., Gjerde, B., Gjedrem, T. & Pullin, R.S.V. (1993). Genetic improvement of farmed tilapias: the growth performance of eight strains of *Oreochromis niloticus* tested in different farm environments. *Aquaculture*, **111**, 171–88.

Fevolden, S.E. & Røed, K.H. (1993). Cortisol and immune characteristics in rainbow trout (*Oncorhynchus mykiss*) selected for high or low tolerance to stress. *Journal of Fish Biology*, **43**, 919–30.

Fevolden, S.E., Refstie, T. & Røed, K.H. (1991). Selection for high and low cortisol stress response in Atlantic salmon (*Salmo salar*) and rainbow trout (*Oncorhynchus mykiss*). *Aquaculture*, **95**, 53–65.

Fevolden, S.E., Refstie, T. & Røed, K.H. (1992). Disease resistance in rainbow trout (*Oncorhynchus mykiss*) selected for stress response. *Aquaculture*, **104**, 19–29.

Fevolden, S.E., Refstie, T. & Gjerde, B. (1993*a*). Genetic and phenotypic parameters for cortisol and glucose stress response in Atlantic salmon and rainbow trout. *Aquaculture*, **118**, 205–16.

Fevolden, S.E., Nordmo, R., Refstie, T. & Røed, K.H. (1993*b*). Disease resistance in Atlantic salmon (*Salmo salar*) selected for high or low responses to stress. *Aquaculture*, **109**, 215–24.

Fevolden, S.E., Røed, K.H. & Gjerde, B. (1994). Genetic components of post-stress cortisol and lysozyme activity in Atlantic salmon; correlations to disease resistance. *Fish and Shellfish Immunology*, **4**, 507–19.

Fjalestad, K.T., Gjedrem, T. & Gjerde, B. (1993). Genetic improvement of disease resistance in fish: an overview. *Aquaculture*, **111**, 65–74.

Gall, G.A.E. (1990). Basis for evaluating breeding plans. *Aquaculture*, **85**, 125–42.

Gall, G.A.E. & Huang, N. (1988*a*). Heritability and selection schemes for rainbow trout: body weight. *Aquaculture*, **73**, 43–57.

Gall, G.A.E. & Huang, N. (1988*b*). Heritability and selection schemes for rainbow trout: female reproductive performance. *Aquaculture*, **73**, 57–66.

Gjedrem, T. (1992). Breeding plans for rainbow trout. *Aquaculture*, **100**, 73–83.

Gjedrem, T. & Gjöen, H.M. (1995). Genetic variation in susceptibility of Atlantic salmon, *Salmo salar* L., to furunculosis, BKD and cold water vibriosis. *Aquaculture Research*, **26**, 129–34.

Gjedrem, T., Salte, R. & Gjöen, H.M. (1991). Genetic variation in susceptibility of Atlantic salmon to furunculosis. *Aquaculture*, **97**, 1–6.

Heath, D.D., Bernier, N.J., Heath, J.W. & Iwama, G.K. (1993). Genetic, environmental, and interaction effects on growth and stress response of chinook salmon (*Oncorhynchus tshawytscha*) fry. *Canadian Journal of Fisheries and Aquatic Sciences*, **50**, 435–42.

Heath, D.D., Bernier, N.J., Heath, J.W. & Iwama, G.K. (1994). Reply to comment on 'Genetic, environmental, and interaction effects on growth and stress response of chinook salmon (*Oncorhynchus tshawytscha*) fry' by Silverstein. *Canadian Journal of Fisheries and Aquatic Sciences*, **51**, 981–3.

Hörstgen-Schwark, G. (1993). Selection experiments for improving 'pan-size' body weight of rainbow trout (*Oncorhynchus mykiss*). *Aquaculture*, **112**, 13–24.

Jones, R.B., Satterlee, D.G. & Ryder, F.H. (1992). Fear and distress in Japanese quail chicks of two lines genetically selected for low or high adrenocortical response to immobilization stress. *Hormones and Behaviour*, **26**, 385–93.

Jones, R.B., Mills, A.D., Faure, J.-M. & Williams, J.B. (1994*a*). Restraint, fear, and distress in Japanese quail genetically selected for long or short tonic immobility reactions. *Physiology and Behaviour*, **56**, 529–34.

Jones, R.B., Satterlee, D.G. & Ryder, F.H. (1994*b*). Fear of humans in Japanese quail selected for low or high adrenocortical response. *Physiology and Behaviour*, **56**, 379–83.

Kinghorn, B. (1983). Genetic variation in food conversion efficiency and growth in rainbow trout. *Aquaculture*, **32**, 141–55.

Lavety, J. (1993). Post-harvest conditions for maximising flesh quality. *Trout News*, **17**, 23–8.

Lowe, T.E., Ryder, J.M., Carragher, J.F. & Wells, R.M.G. (1993). Flesh quality in snapper, *Pagrus auratus*, affected by capture stress. *Journal of Food Science*, **58**, 770–3.

Mason, D. (1991). Genetic variation in the stress response: susceptibility to experimental allergic encephalomyelitis and implications for human inflammatory disease. *Immunology Today*, **12**, 57–60.

Meaney, M.J., Bhatnagar, S., Diorio, J., Larocque, S., Francis, D., O'Donnell, D., Shanks, N., Sharma, S., Smythe, J. & Viau, V. (1993). Molecular basis for the development of individual differences in the hypothalamic–pituitary–adrenal stress response. *Cellular and Molecular Neurobiology*, **13**, 321–47.

McGeer, J.C., Baranyi, L. & Iwama, G.K. (1991). Physiological responses to challenge tests in six stocks of coho salmon (*Oncorhynchus kisutch*). *Canadian Journal of Fisheries and Aquatic Sciences*, **48**, 1761–71.

Möck, A. & Peters, G. (1990). Lysozyme activity in rainbow trout, *Oncorhynchus mykiss* (Walbaum), stressed by handling, transport and water pollution. *Journal of Fish Biology*, **37**, 873–85.

Nilsson, J. (1992). Genetic variation in resistance of Arctic char to fungal infection. *Journal of Aquatic Animal Health*, **4**, 126–8.

Pickering, A.D. (1993). Growth and stress in fish production. *Aquaculture*, **111**, 51–63.

Pickering, A.D. (1984). Cortisol-induced lymphocytopenia in brown trout, *Salmo trutta* L. *General and Comparative Endocrinology*, **53**, 252–9.

Pottinger, T.G. & Moran, T.A. (1993). Differences in plasma cortisol and cortisone dynamics during stress in two strains of rainbow trout (*Oncorhynchus mykiss*). *Journal of Fish Biology*, **43**, 121–30.

Pickering, A.D. & Pottinger, T.G. (1987). Poor water quality suppresses the cortisol response of salmonid fish to handling and confinement. *Journal of Fish Biology*, **30**, 363–74.

Pickering, A.D. & Pottinger, T.G. (1989). Stress responses and disease resistance in salmonid fish: effects of chronic elevation of plasma cortisol. *Fish Physiology and Biochemistry*, **7**, 253–8.

Pottinger, T.G., Pickering, A.D. & Hurley, M.A. (1992). Consistency in the stress response of individuals of two strains of rainbow trout, *Oncorhynchus mykiss*. *Aquaculture*, **103**, 275–89.

Pottinger, T.G., Moran, T.A. & Morgan, J.A.W. (1994). Primary and secondary indices of stress in the progeny of rainbow trout (*Oncorhynchus mykiss*) selected for high and low responsiveness to stress. *Journal of Fish Biology*, **44**, 149–63.

Pottinger, T.G., Balm, P.H.M. & Pickering, A.D. (1995). Sexual maturity modifies the responsiveness of the pituitary–interrenal axis to stress in male rainbow trout. *General and Comparative Endocrinology*, **98**, 311–20.

Price, E.O. (1984). Behavioural aspects of animal domestication. *Quarterly Review of Biology*, **59**, 1–32.

Purdom, C.E. (1993). *Genetics and Fish Breeding*. Chapman & Hall, London.

Rand-Weaver, M., Pottinger, T.G. & Sumpter, J.P. (1993). Plasma somatolactin concentrations in salmonid fish are elevated by stress. *Journal of Endocrinology*, **138**, 509–15.

Refstie, T. (1982). Preliminary results: differences between rainbow trout families in resistance against vibriosis and stress. *Developmental and Comparative Immunology Supplement*, **2**, 205–9.

Refstie, T. (1986). Genetic differences in stress response in Atlantic salmon and rainbow trout. *Aquaculture*, **57**, 374.

Refstie, T. (1990). Application of breeding schemes. *Aquaculture*, **85**, 163–9.

Røed, K.H., Larsen, H.J.S., Linder, R.D. & Refstie, T. (1993). Genetic variation in lysozyme activity in rainbow trout (*Oncorhynchus mykiss*). *Aquaculture*, **109**, 237–44.

Rye, M., Lillevik, K.M. & Gjerde, B. (1990). Survival in early life of Atlantic salmon and rainbow trout: estimates of heritabilities and genetic correlations. *Aquaculture*, **89**, 209–16.

Salonius, K. and Iwama, G.K. (1991). The effect of stress on the immune function of wild and hatchery coho salmon juveniles. *Bulletin of the Aquaculture Association of Canada*, **91–3**, 47–9.

Salonius, K. & Iwama, G.K. (1993). Effects of early rearing environment on stress response, immune function, and disease resistance in juvenile coho (*Oncorhynchus kisutch*) and chinook salmon (*O. tshawytscha*). *Canadian Journal of Fisheries and Aquatic Sciences*, **50**, 759–66.

Satterlee, D.G. & Johnson, W.A. (1988). Selection of Japanese quail for contrasting blood corticosterone response to immobilization. *Poultry Science*, **67**, 25–32.

Silverstein, J.T. (1994). Comment on 'Genetic, environmental, and interaction effects on growth and stress response of chinook salmon (*Oncorhynchus tshawytscha*) fry' by Heath *et al.* (1993). *Canadian Journal of Fisheries and Aquatic Sciences*, **51**, 981.

Sumpter, J.P., Pickering, A.D. & Pottinger, T.G. (1985). Stress-induced elevation of plasma α-MSH and endorphin in brown trout, *Salmo trutta* L. *General and Comparative Endocrinology*, **59**, 257–65.

Tave, D. (1993). *Genetics For Fish Hatchery Managers*, Second Edition. Van Nostrand Reinhold, New York.

Tave, D., Smitherman, R.O. & Jayaprakas, V. (1989). Estimates of additive genetic effects, maternal effects, specific combining ability, maternal heterosis, and egg cytoplasm effects for cold tolerance in *Oreochromis niloticus* (L.). *Aquaculture and Fisheries Management*, **20**, 159–66.

Taylor, J., Weyers, P., Harris, N. & Vogel, W.H. (1989). The plasma catecholamine stress response is characteristic for a given animal over a one-year period. *Physiology and Behaviour*, **46**, 853–56.

Thorpe, J.E., Morgan, R.I.G., Talbot, C. & Miles, M.S. (1983). Inheritance of developmental rates in Atlantic salmon, *Salmo salar* L. *Aquaculture*, **33**, 119–28.

Withler, R.E. & Beacham, T.D. (1994). Genetic variation in body weight and flesh colour of the coho salmon in British Columbia. *Aquaculture*, **119**, 135–48.

Woodward, C.C. and Strange, R.J. (1987). Physiological stress responses in wild and hatchery-reared rainbow trout. *Transactions of the American Fisheries Society*, **116**, 574–9.

Yousif, A.N., Albright, L.J. & Evelyn, T.P.T. (1994). *In vitro* evidence for the antibacterial role of lysozyme in salmonid eggs. *Diseases of Aquatic Organisms*, **19**, 15–19.

P.H.M. BALM

Immune–endocrine interactions

Introduction

Fish mobilize defence systems to counteract the negative effects of noxious stimuli. An important regulatory role in these events is played by the (neuro-)endocrine system (see chapters by B.A. Barton and J.P. Sumpter, this volume). When external stimuli are experienced as stressful, the activation of the hypothalamus–pituitary–interrenal (HPI) axis is considered to initiate the organism's adaptive response (Donaldson, 1981; Pickering, 1989). The stress concept (originally introduced by Selye) has been generalized recently to include the response to a wide array of influences, including parasitic and bacterial infections (Plytycz & Seljelid, 1995). Natural defence against these insults is provided by the host's immune system, which consequently has been considered part of the animal's adaptive repertoire (Bayne, 1994).

The present paper reviews the recent evidence that regulatory systems such as the neuroendocrine and the immune systems do not operate in isolation in reaction to specific stimuli, but act in a concerted way, thereby increasing the organism's capacity to cope with a suprisingly variable collection of stimuli or stressors threatening vital functions such as growth and reproduction (DeKloet & Voorhuis, 1992). These novel insights into adaptive regulation most likely have aspects which are applicable to, and which should increase our understanding of, phenomena associated with aquaculture practices such as: (1) outbreaks of disease (Snieszko, 1974; G.A. Wedemeyer, this volume), which prevent a significant portion of fish from reaching a marketable size; (2) the fact that preventive treatments such as vaccination have been demonstrated to be less effective in stressed fish; and (3) the development of autoimmune reactions, one of the negative aspects of rearing fish at high density (Stickney, 1986). Realization of the integration between the neuroendocrine and immune systems in fish will allow fish culturists to assess better the implications of immunotherapeutic or

stressful treatments for fish performance, which should lead to optimiz-
ation of stock management. The fundamental impact of neuroimmunological [1] principles is furthermore indicated by studies demonstrating
that they do not operate exclusively in stressed and diseased states,
but also during 'routine' functions such as exercise, development, intermediate metabolism and ion regulation (Cooke, 1994; Hoffman-
Goetz & Pedersen, 1994; Maule, Schreck & Fitzpatrick, 1994).

The current concept of concerted actions between neuroendocrine
and immune systems was originally described in relation to mammals
(Besedovsky & Sorkin, 1977; Blalock, 1989), although evidence that
these systems also communicate during conditions of stress in fish has
been available for several decades (Rasquin, 1951). In higher vertebrates, evidence is now overwhelming that the bidirectional communication between the neuroendocrine and the immune systems operates at
all levels of organization (organism, organ, cell). Figure 1 schematically
illustrates this concept, whereby the biological response of an organism
to external stimuli is determined to a large extent through interactions
between regulatory systems. This communication is facilitated by the
action of messengers and receptors shared by regulatory systems
('common messengers and receptors' hypothesis; Blalock, 1989), which,
until recently, were considered to operate independently. The immune
and the central nervous systems serve as excellent examples of that
category, with cytokines (first described as molecules belonging to the
immune system), such as interleukin 1 (IL-1) and tumour necrosis
factor alpha (TNF-α), or neuropeptides (traditional neuroendocrine
molecules), serving as multifunctional messengers, active in both systems. This rapidly expanding field of research has been reviewed
recently (Millington & Buckingham, 1992; Gaillard, 1994; Lotan &
Schwartz, 1994; Madden, Sanders & Felten, 1995; Rivier, 1995;
Rothwell & Hopkins, 1995; Wilder, 1995) and I will deal here with
aspects that bear directly on current developments in immune–endocrine research in fish.

Neuroimmunological research developed from research into the regulation of the hypothalamus–pituitary–adrenal (HPA) axis in animals
challenged with infectious agents (Besedovsky & Sorkin, 1977), which
illustrates the fundamental relationship between stress and disease. The
HPA axis appears to be regulated by cytokines, such as IL-1, TNF-α,
IL-6, thymosin V (Gaillard, 1994; Rivier, 1995).

---

[1]In the mammalian literature, the term neuroimmunology covers the broad field of research into the
interactions between the (neuro-)endocrine and immune systems. In relation to fish, the more general
'immune–endocrine interactions' may be more appropriate. It puts less emphasis on neural influences,
which, to date, have been studied scantily in relation to immune function in fish.

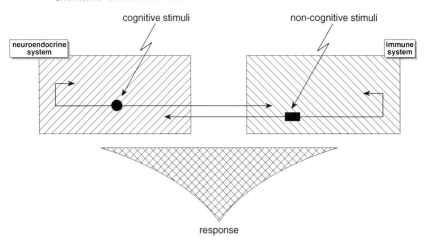

Fig. 1. Schematic illustration of the concept of bidirectional communication between the endocrine and the immune system. A stimulus perceived by one of the systems is not processed solely within that particular system, but also exerts its influence in the other system (horizontal arrows). This ensures that challenges perceived by the central nervous system (CNS) or the immune system are translated into an appropriately integrated response. Communication between systems is facilitated by the presence of common messengers (hormones, cytokines) and receptors. This concept was first formulated for mammals by Besedovsky and Sorkin (1977) and Blalock (1989), and was based on findings demonstrating that: (1) immune responses can be conditioned, (2) stressed animals have altered immune responses and altered susceptibility to tumours, and (3) activation of the immune system can be correlated with altered neurophysiological, neurochemical and neuroendocrine activities of brain cells.

The finding that the synthesis of the multifactorial precursor protein proopiomelanocortin (POMC) is not confined to the neuroendocrine system, but also takes place in lymphoid cells (Blalock, 1989; Lyons & Blalock, 1995), further illustrates the integration between the immune and neuroendocrine systems. Subpopulations of lymphoid cells also express receptors for some of these peptides. POMC, among other peptides, encodes for adrenocorticotropin (ACTH) and endorphin, first identified as key hormones of the HPA axis. The special role of POMC in immune–endocrine regulation is further emphasized by the actions of another hormone from this precursor, α-melanocyte stimulating hormone (α-MSH), in the regulation of immune functions (Catania & Lipton, 1993). A major experimental difficulty in establishing immuno-modulatory actions of peptide hormones (Sharp & Linner, 1993) is that expression of both peptides and their receptors generally appears to be low in cells from the immune system. The recent demonstration of the presence of PC1 and PC2, two POMC processing enzymes, in neutrophils and macrophages of the rat, however, further substantiates the presence and function of endogenous 'neuro' peptides in the immune system (Vindrola et al., 1994).

Immune–endocrine interactions may hold the key to understanding previously suspected interactions among stress, health and disease. In a series of contributions, Sternberg and coworkers reported that the susceptibility of female Lewis rats to streptococcal cell wall components (this rat strain displays an increased incidence of arthritis when treated with these preparations) was caused by a defect in the adrenal activation that occurs in inflammatory-disease-resistant Fischer rats receiving the bacterial preparation. Apparently, during inflammation the activation of the pituitary–adrenal axis, through a rise in circulating immunosuppressive corticosteroids, prevents the immune system from 'overshooting'. Proving that neuroimmunological regulation is critical in these cases, it was demonstrated that the absence of an adrenal gland activation in the Lewis rats was caused by the insensitivity to IL-1 of their hypothalamic neurosecretory cells producing corticotropin-releasing hormone (CRH) (Sternberg et al., 1989).

More recently, it has become clear that immunoendocrine signalling is not restricted to the HPA axis, but that it probably governs a large number of tissues, including perhaps less evident ones such as the skin (Bhardwaj & Luger, 1994; Sim, 1995) and endothelium (Oswald et al., 1994). Along with the growing list of identified hormones produced by the haematopoietic system (and their functions; Hooghe et al., 1993), increasing numbers of physiological actions of cytokines are being discovered. The function of sessile macrophages in multiple organs and their interplay with hormones in the regulation of ovarian and testicular function (Hutson, 1994) have led to the assumption that an important part of the neuroimmunological interactions probably takes place via paracrine communication. As a result of immune–endocrine research, the distinctions between the immune and the endocrine systems therefore become increasingly vague, in particular at the paracrine level. The current integrated knowledge of the mechanisms behind the physiological and endocrinological consequences of bacterial infections corroborates the contention that immune–endocrine signalling probably takes place in most, if not all, organs. Important tools in this research are lipopolysaccharides (LPS), cell wall components of Gram-negative bacteria. These molecules are responsible for septic shock (hence the term 'endotoxin'), through a cascade of cytokine-mediated actions (Tilders et al., 1994).

From a biomedical point of view, the importance of neuroimmunological integration is not confined to bacterial infections, but also concerns viral infections (Sundar *et al.*, 1991). The link with autoimmune diseases such as rheumatoid arthritis has been hinted at (Sternberg *et al.*, 1989), but it is now evident that disorders associated with sleep, behaviour, metabolism and addiction have neuroimmunological aspects. Antibodies against TNF-α have been suggested as therapeutic agents against septic shock.

It is probable that immune–endocrine mechanisms developed early in evolution, since the principles of neuroimmunolgy have been demonstrated in invertebrates (Scharrer, 1991; Ottaviani, Franchini & Fontanili, 1994; Ottaviani, Capriglione & Franceschi, 1995*b*), although the 'ingredients' are partly specific to these animals (Ottaviani, Caselgrandi & Franceschi, 1995*a*). Here also, these regulatory processes appear to play pivotal roles in the animals performance when exposed to pathogens such as schistosomes (de Jong-Brink, 1995). Studies of invertebrates have strengthened the view that the immune system, including the non-lymphoid components, is basically part of the animals' adaptive repertoire (Bayne, 1994), and have established the importance of a comparative approach when interpreting immune–endocrine regulations (Scharrer, 1991).

From what has been written so far, it might be anticipated that immune–endocrine communication may also take place in bony fishes. This suggestion is supported by an increasing body of experimental evidence, which will be dealt with in the following sections. At present, this evidence stems mainly from fundamental research, and application of the concept to aquaculture will most likely follow from further study. The basic aspects of immune–endocrine integration in fish will be discussed, using the framework provided by mammalian models (which will be referred to incidentally from hereon). Due to space limitations, it will be impossible to acquaint the reader with the fish immune system in this chapter, but basic information can be found in several recent reviews (Rowley *et al.*, 1988; Kaattari, 1992; Secombes & Fletcher, 1992; Pohajdak, Dixon & Stuart, 1993; Secombes, 1994). Importantly, fish were the first animals to possess lymphocytes and to have truly acquired immune functions.

In retrospect, evidence for immune–endocrine interactions in teleost fish has been available for more than 40 years (Rasquin, 1951), but the majority of data stems from recent studies, and progress since the first general paper (Ndoye *et al.*, 1991) has been considerable. The remaining part of this chapter will consist of three sections:

1. the influence of hormones on the immune system of fish

2. regulation of the endocrine system by components of the immune system; effects of LPS and
3. aspects dealing with health and disease; future prospects.

The influence of hormones on immune functions in fish

From the initial studies performed with fish, it became clear that modulation of endocrine functions consequently altered the activity of immune components. As a result of experimental data demonstrating immunosuppressive consequences of an activation of the HPI axis (Rasquin, 1951), attention focused for quite some time (almost exclusively) on the effects of corticosteroids; the severe furunculosis experienced by spawning salmonids for instance was related primarily or in fact solely, to the concurrent interrenal hypertrophy (Donaldson, 1981). The relationship between environmental stressors and disease resistance (Wedemeyer, 1970; Snieszko, 1974; Fries 1986) appeared to be strongly dependent on activation of interrenal cortisol output, because exogenous administration of ACTH (McLeay, 1973), cortisol (Pickering, 1984), as well as experimentally applied stressful conditions (Pickford et al., 1971) resulted in modulation of immune functions. As a consequence it became evident that stress and, in particular, cortisol surges associated with aquaculture practices, such as crowding and/or transportation, result in fish becoming more susceptible to infestations by parasites and other microorganisms (Walters & Plumb, 1980; Maule et al., 1989; Barton & Iwama, 1991). Paradoxically, some of the preventive treatments may further contribute to increased susceptibility. Treatment with products such as kanamycin, formalin and malachite green (Donaldson, 1981) have been demonstrated to result in elevated levels of cortisol, an immunosuppressive hormone.

Despite the initial focus on the immunosuppressive effects of stress, more recently it has become apparent that the effects of stress on the fish immune system are varied and complex, and depend on the parameter and/or tissue studied, and on the applied stressor. Thompson et al. (1993) demonstrated that stress lowered the bactericidal activity of tissue leucocytes from Salmo salar, but increased the plasma bactericidal activity. Also the effects of cortisol (usually considered as immunosuppressive) have been found to be more varied than previously realized, in particular when the mechanisms of action at the whole-animal level are considered (Maule, Schreck & Sharpe, 1993). These mechanisms governing immune–endocrine communication might also differ between fish species (and clearly between fish and mammals!). The reported (Iger et al., 1995) increased rate of leucocyte apoptosis in the skin of rainbow trout treated with exogenous cortisol, for instance, ties in with the effects of cortisol on lymphocyte apoptosis reported in relation to mammals. Conversely, Alford *et al.* (1994) reported a decrease in leucocyte apoptosis in stressed fish, and failed to demonstrate an *in vitro* effect of cortisol on the same parameter.

The complexity of the interrelationship between the immune and the endocrine systems is well illustrated by the effects of social stressors. Ghoneum *et al.* (1988) demonstrated a rapidly reduced cell-mediated immunity in tilapia hybrids confronted with aggressive siblings. In that study, β specimens (subordinates) were invaded more by internally or externally applied *Aeromonas* compared with non-stressed individuals. Phagocytosis in rainbow trout (*Oncorhynchus mykiss*) can be stimulated or inhibited in subordinates (Peters *et al.*, 1991), depending mainly on the acute or chronic nature of the confrontation. The work of Faisal and coworkers (1989, 1992) sheds some light on the immune–endocrine regulation behind these effects, demonstrating β-endorphin (β-END) and cholinergic modulation of immune functions under these circumstances, in particular in subordinate fish.

The complexity of the effects of stress on immune function is demonstrated by the multihormonal nature of the interactions. Table 1 lists data available on the effects of hormones (either administered *in vivo* or *in vitro*) on immunity parameters. Particularly for the effects of cortisol (and dexamethasone), this table should not be regarded as exhaustive; studies of these hormones have been summarized by others recently (Pickering, 1989; Barton & Iwama, 1991). Neither does the list mention studies that indirectly implicate immunoregulatory actions of hormones, such as thyroid hormones (Ball & Hawkins, 1976), growth hormone (Marc *et al.*, 1995), or prolactin (Betoulle *et al.*, 1995).

Several aspects become apparent from the research summarized in Table 1.

- As in other areas of fish research, studies with rainbow trout are dominant. For a general review, therefore, it seems not to be appropriate to draw major conclusions or to make general comparisons on the basis of the evidence available. Undoubtedly this list will grow, and the need for data on species other than rainbow trout is indicated.
- Despite the relatively limited amount of data, several discrepancies are evident, which may partly relate to the doses employed (Pickering, 1984). The different effects on macrophages observed with α-adrenergic agonists *in vitro* suggest that different activities (phagocytosis and superoxide

Table 1. *Effects of hormones administered in vivo (Table 1a) or in vitro (Table 1b) on immunity parameters*

a **in vivo**

Factor	Species	Parameter	Effect	Reference
Cortisol	*Salmo trutta*	Circulating lymphocytes	Reduction	Pickering (1984)
	Oncorhynchus kisutch	Frequency B cell precursors	Decrease	Kaattari & Tripp (1987)
	Ictalurus punctatus	(1) Circulating lymphocytes; (2) circulating neutrophils; (3) lymphocyte responsiveness to mitogens	(1) Decrease; (2) increase; (3) decrease	Ellsaesser & Clem (1987)
	Oncorhynchus kisutch	Generation of antibody-producing cells	Reduction	Maule *et al.* (1987)
	Oncorhynchus masou	Plasma immunoglobulin M	Suppression	Nagae *et al.* (1994)
	Oncorhynchus mykiss	Leucocyte migration/ apoptosis skin	Enhancement	Iger *et al.* (1995)
Oestradiol	*Carrasius auratus*	Mitogen-induced proliferation of peripheral blood lymphocytes	Suppression	Wang & Belosevic (1994)
Naltrexone (opioid antagonist)	*Tilapia mossambica × Tilapia honorum*	Cytotoxicity/ proliferation of nephric leucocytes	Prevention of drop in cytotoxicity during aggressive encounters, as well as the drop in mitogen responsiveness; leucocyte proliferation	Faisal *et al.* (1989)

b **in vitro**

Factor	Species	Parameter	Effect	Reference
Cortisol	Morone saxatilis	Response of phagocytes to Aeromonas hydrophila	Inhibition of chemiluminescence	Stave & Robertson (1985)
	Oncorhynchus kisutch	B cell activation (plaque-forming response)	Inhibition (lymphokine-like factor involved)	Tripp et al. (1987)
	Carassius auratus	Nitric oxide production macrophage cell line	Inhibition	Wang & Belosevic (1995)
Testosterone	Oncorhynchus tshawytscha (juveniles)	Antibody-producing cells	Suppression of plaque-forming response (season dependent)	Slater & Schreck (1993)
Oestradiol	Carassius auratus	Proliferation of peripheral blood lymphocytes	Suppression	Wang & Belosevic (1994)
	Carassius auratus	Chemotaxis and phagocytosis macrophage cell line	Inhibition	Wang & Belosevic (1995)
ACTH	Oncorhynchus mykiss	Respiratory burst; phagocytotic leucocytes	Stimulation	Bayne & Levy (1991)
Carbachol (cholinergic agonist)	Oncorhynchus mykiss	Antibody response to TNP-LPS	Stimulation	Flory (1990)
Endorphin	Tilapia mossambica × Tilapia honorum	Cytotoxicity/proliferation of pronephric leucocytes	Suppression (reversible by naltrexone)	Faisal et al. (1989)

Table 1.b *in vitro*—cont.

Factor	Species	Parameter	Effect	Reference
Adrenergic agents	*Oncorhynchus mykiss*	Respiratory burst	Suppression (adrenaline and β-adrenergic agonist)	Bayne & Levy (1991)
	Oncorhynchus mykiss	Antibody response to TNP-LPS	Suppression by isoproterenol (β-adrenergic agonist), stimulation by clonidine (α_2-adrenergic agonist)	Flory (1990)
	Oncorhynchus mykiss	Phagocytotic activity of macrophages (pronephros and spleen)	Suppression by phenylephrine and isoprenaline (α- and β-adrenergic agonists)	Narnaware, Baker & Tomlinson (1994)
Prolactin, growth hormone	*Oncorhynchus mykiss*	Superoxide anion production by macrophages	Stimulation	Sakai, Kobayashi & Kawauchi (1995)
Growth hormone	*Sparus aurata*	Proliferation of head kidney cells, enriched in leucocytes	Stimulation (mimicked by IGF-1)	Calduch-Giner et al. (1995)
Somatostatin	*Oncorhynchus mykiss*	Proliferation of peripheral blood leucocytes	Suppression (basal and in particular LPS stimulated)	Ndoye et al. (1991)
Substance P	*Oncorhynchus mykiss*	Proliferation of peripheral blood leucocytes	Stimulation (basal and in particular PHA stimulated)	Ndoye et al. (1991)

production) of the same cell type do not necessarily respond in the same direction to a stimulus (Narnaware, Baker & Tomlinson, 1994).

- Cells obtained from different tissues display varying sensitivities to treatments (Tripp *et al.*, 1987).

Although Faisal *et al.* (1989) demonstrated that the immunosuppression observed to occur in β fish could be related to a circulating endorphin-like factor, it is likely that many of these mechanisms *in vivo* will operate at the paracrine (local) level of organization, rather than at the level of the whole organism. In relation to this, it is important to consider that the hormones involved most likely also include those produced in the immune system itself, such as ACTH or growth hormone. These are not necessarily identical to the pituitary-derived counterparts, which are generally applied to fish (Bayne & Levy, 1991; Calduch-Giner *et al.*, 1995). This aspect of immune–endocrine integration in fish is hinted at in a few papers (Balm *et al.*, 1995*a*; Ottaviani *et al.*, 1995*b*) only, and solely concerns immunoreactive fractions, which await characterization. Despite this incomplete picture, it seems established beyond reasonable doubt that the endocrine system is intimately interwoven with the immune system in fish. The recent finding that fish selected for interrenal responsiveness to stressors (Fevolden, Refstie & Røed, 1992) also display characteristic differences in immune functions strengthens this view (see chapter by T.G. Pottinger and A.D. Pickering, this volume). Apart from the recent characterization of an androgen receptor in rainbow trout lymphocytes (Slater, Fitzpatrick & Schreck, 1995) and partial characterization of other receptors involved, virtually nothing is known about the mechanisms behind the humoral effects on immune functions in fish; results obtained by Tripp *et al.* (1987) and Kaattari and Tripp (1987) point to the involvement of an IL-1-like factor in the cortisol-induced suppression of the number of antibody-producing cells in salmon.

Regulation of the endocrine system by immune components; effects of LPS

There is also a paucity of information concerning this direction of the bidirectional communication between the two systems. However, the data available serve as reliable indications of its operation. It has been recognized for some time that conditions which alter immune parameters often are accompanied by changes in the activity of endocrine tissues. Most data concern the HPI axis, which has been demonstrated to be activated in fish experiencing a wide range of infestations by microorganisms (see Donaldson, 1981). However, this activation has been suggested not to be typical for fish under these circumstances, in contrast to the situation in mammals. Rand and Cone (1990), for instance, failed to demonstrate elevated cortisol levels in trout experimentally infected with *Ichthyophonus hoferi*.

As discussed in the first part of this chapter, identification of the messenger molecules utilized in the communication with the endocrine system (cytokines, such as IL-1) has greatly furthered research into the mechanisms of endocrine–immune interactions in mammals. However, despite considerable effort by several groups, positive identification of fish 'cytokines' is restricted to macrophage-activating factor (MAF), the teleost homologue of γ-interferon (Graham & Secombes, 1988; Francis & Ellis, 1994). Nevertheless, there are several indications that fish might produce a fair amount of cytokines and, although extensive discussion of the literature on this point falls beyond the scope of this chapter, it will be reviewed briefly, as the identification of homologous cytokines will facilitate future developments in immune–endocrine research in fish. Probably the most sought after factor is the teleost equivalent of IL-1, and several papers show the likely existence of a teleost IL-1-like factor (Hamby *et al.*, 1986; Kaattari & Tripp, 1987; Ellsaesser & Clem, 1994). Ghanmi, Rouabhia and Deschaux (1993) recently provided evidence for the presence of IL-2 in carp. Although the presence of TNF-α, another important cytokine in immune–endocrine communication in mammals, has not been demonstrated yet, recombinant TNF exerts effects on fish immune functions, which can be blocked by antisera directed against the mammalian TNF-α receptor (Jang *et al.*, 1995). More extensive is the evidence for transforming growth factor beta (TGF-β) in teleosts (Jang, Hardie & Secombes, 1994). Although Hausmann (1995) recently cautioned against presumptuous 'identifications' of mammalian-like cytokines in fish, the data available presently support the idea of an interactive cytokine network regulating immune responses in fish, which may also exert regulatory influences on the endocrine system.

Experimental evidence for this latter suggestion is limited to several papers at present. Schreck and Bradford (1990) demonstrated that coho salmon leucocytes in culture release a factor(s) which influences cortisol production by interrenal cells *in vitro*. Balm *et al.* (1993) described the inhibitory effects of *in vivo* administered recombinant murine IL-1 on pituitary α-MSH release, both unstimulated and in response to thyrotropin-releasing hormone (TRH), in tilapia (*Oreochromis mossambicus*). Figure 2 illustrates more recent data on the *in vitro* effects of TNF-α on the same parameter. In mammals, the effects of

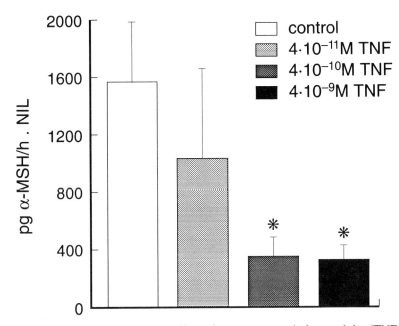

Fig. 2. Dose-dependent effect of tumour necrosis factor alpha (TNF-α, human recombinant, Genzyme) on alpha-melanocyte-stimulating hormone (α-MSH) release from tilapia (*Oreochromis mossambicus*) pituitary neurointermediate lobes (NIL) incubated statically *in vitro* for 2 h. Data are expressed as averages ± SEM; $n = 5$ for all groups. For details on experimental procedures, see Balm *et al.*, 1993.

TNF-α and IL-1 overlap to a high degree. Because the inhibitition of α-MSH release by TNF-α in tilapia (Fig. 2) resembles the IL-1 effect reported by Balm *et al.* (1993), the same may hold for the teleost equivalents of IL-1 and TNF-α. Studying the effects of LPS on endocrine tissues appears to be an experimental approach that does not require homologous factors, but which strongly substantiates the operation of immune–endocrine signalling. As in mammals, this preparation from Gram-negative bacterial cell walls has been known for some time to affect the HPI axis *in vivo* in fish, generally in a stimulatory fashion (Wedemeyer, 1969; White & Fletcher, 1985; Balm *et al.*, 1995a). Because these effects, as in higher vertebrates (Tilders *et al.*, 1994), are most likely to be mediated through components of the immune system, the data available support the idea that the link between the immune system and the pituitary–adrenal (–interrenal) axis is a phylogenetically old one.

Because research into the mechanisms underlying immune–endocrine interactions at present is hampered by the lack of knowledge about homologous cytokines (and hormones for that matter!), it is not clear to what extent the picture in fish will resemble, or differ from, that in mammals. Based on the morphology of several tissues that are likely to stage immune–endocrine interactions, it is, however, safe to assume that there will be many fish-specific aspects, which will be discussed next.

The high incidence of macrophage-like cells in the teleost central nervous system (Dowding & Scholes, 1993), which – in contrast to mammalian CNS cells of the macrophage/monocyte lineage – appear to migrate between the CNS and the periphery, points to potentially novel mechanisms of immune–endocrine signalling. In particular, the question of how information regarding the status of the immune system is conveyed to the CNS may be easier to answer in fish than in mammals. However, recently it was suggested that haematopoietic cells may also play a role in neural–endocrine interactions in the mammalian CNS (Silver et al., 1996).

Another striking example of differences between fish and mammalian immune–endocrine regulations concerns the head kidney, which contains the interrenal cortisol-producing cells. In fish, the head kidney is also one of the most important haematopoietic tissues, in which no apparent boundaries exist between immune and endocrine cells. The possible functional significance of this organization, unique to fish, is documented in studies by Schreck and Bradford (1990) and Balm et al. (1995a). The latter study demonstrated that treatment of head kidneys of tilapia with LPS in vitro changed the ACTH responsiveness of the interrenal cells (Fig. 3). LPS treatment also provoked the release of an α-MSH immunoreactive factor from the tissue, which may be implicated in the regulation of ACTH responsiveness.

Other promising models for identifying teleost-specific interactions between the endocrine and immune systems are the skin and in particular, the gills. In an environment rich in microorganisms, the effectiveness of these organs to counteract challenges by pathogens determines, to a large extent, the susceptibility of fish to diseases. The stress-related modulation of disease susceptibility may also take place at these levels since, to a large extent, the animal's defence is organized extracellularly by humoral factors (lysozyme, immunoglobulin G), as well as directly cell mediated in the epidermis/branchial epithelium. The latter is indicated by tissue leucocyte mobilization/infiltration when fish are challenged, or when an increased immune surveillance is called for (Balm & Pottinger, 1993). The vulnerability of the gills in particular is intimately

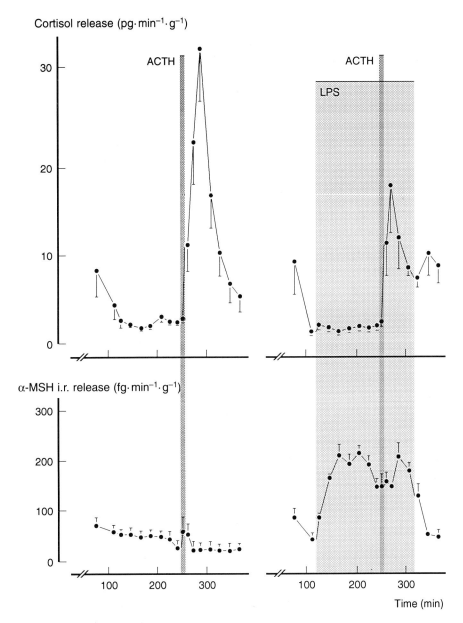

Fig. 3. *In vitro* effects of LPS (*Escherichia coli* lipopolysaccharide) on spontaneous and adrenocorticotropin-(ACTH-)stimulated cortisol and α-MSH-like immunoreactivity release by head kidneys from tilapia (*Oreochromis mossambicus*). Reprinted with permission from Balm *et al.* (1995a).

connected to its morphology, which is designed to allow physiological functions such as gas exchange and ion transport. These functions are under hormonal control, but it is conceivable that several of these branchial functions are co-regulated by, or involve, immunocytes as well. A comparable development in mammalian research has been seen recently in mammalian research (Cooke, 1994), where it appears that immune, nervous and epithelial systems are intimately linked, providing the organism with integrated responses to physiological and pathophysiological stimuli. Indications that the same might also occur in fish gills stem from data demonstrating that the proliferation of chloride cells (CC; cells in the branchial epithelium specialized for ion translocation) can be manipulated experimentally by the *in vivo* administration of recombinant murine IL-1 or *Escherichia coli* LPS (Balm *et al.*, 1995a). The IL-1 effect occurred independently of interrenal activation, which makes it likely that signals from tissue leucocytes served as intermediates for the CC proliferation. Similar regulatory mechanisms may be responsible for branchial effects observed to occur in rainbow trout experimentally infected with *Flavobacterium branchiophilum*, the causative agent of bacterial gill disease (BGD) (Byrne *et al.*, 1995). Apparently, the initial pathogenesis of BGD is not related to non-specific tissue damage, but largely results from functional impairment of branchial physiological processes, which, most likely, are chemically mediated. Recent evidence suggests that ion transport mechanisms located in branchial CC might be direct targets for toxins produced by microorganisms (Gaete *et al.*, 1994). Finally, immune influences on gill function may also be mediated by the extensive diffuse 'neuro'–endocrine system in this tissue (Zaccone, Fasulo & Ainis, 1995).

Skin resembles the gills in many aspects, particularly when it comes to the immune surveillance carried out by immunocytes in the epidermis. This surveillance is intensified very rapidly when fish are confronted by a wide variety of challenges (Iger *et al.*, 1995; and references therein) not necessarily aimed primarily at the immune system. Cellular invasions, but also an increase in lysozyme contents, and changes in mucus-producing cells (Ferguson *et al.*, 1992) are among the responses to these challenges. Interestingly, many of the aspects of epidermal immune activations can be mimicked by cortisol administration under stress-free conditions (Iger *et al.*, 1995). This corroborates the idea that the endocrine and the immune systems also interact in peripheral tissues, such as the skin. Future research will have to prove whether, in fish, 'hormones' endogenously produced by such tissues play a role in these interactions. Recently it was established that the mammalian skin contains cells producing POMC, which is posttranslationally modified to factors such as α-MSH (Bhardwaj & Luger, 1994). To date, research into the role of this hormone in fish has almost exclusively focused on the actions of pituitary-derived α-MSH in processes such as background adaptation (Bowley, Rance & Baker, 1983). The most pronounced activation of these pituitary α-MSH-producing cells occurs when certain species of fish are placed on a black background. Several indications, however, link α-MSH with immune functions. Keeping fish on a white versus a black background also influences those functions of skin related to defence against microorganisms. In 1938, Sumner and Douderoff observed that the background colour influenced disease resistance in fish, whereas more recently it was reported that black-adapted (BA) fish skin is richer in superoxide dismutase (SOD) compared with the skin of counterparts adapted to a white background (Nakano, Sato & Takeuchi, 1995). Moreover, heavily infected fish generally turn black, which may or may not be related to the finding of Bowley *et al.* (1983) that plasma α-MSH levels increased in fungally infected fish. Finally, the action of α-MSH to increase melanin biosynthesis in the skin might directly or indirectly function in the animal's defence mechanism (discussed in Balm *et al.*, 1995*b*). Synthesis takes place in the dermal melanophores but, when challenged, fish melanophore extensions invade the epidermis and melanin is transferred to epidermal macrophages (Iger *et al.*, 1995), which bear morphological resemblance to melano-macrophage centres (Martinez *et al.*, 1994).

Therefore, there seem to be more than enough indications of interplay between the endocrine and immune systems in the integument of fish in a wide variety of physiological functions not necessarily restricted to diseased states.

Health and disease; future prospects

This section deals briefly with future directions that may substantiate the fundamental impact of immune–endocrine communication in fish. Three areas need special attention. First, it is imperative that information becomes available on the character of homologous fish cytokines and hormones. Data are scarce, in particular regarding messenger molecules of the immune system. It may well turn out that heterologous molecules can be used in fish immune–endocrine research but, until this is clear, many investigators will refrain from investing much time and effort in this area.

A second approach that will further the recognition of immune–endocrine communication in fish is the incorporation of immunological aspects in the regulation of reproduction. In higher vertebrates, for instance, it has been established that gonadal sessile macrophages are important regulators of reproductive function (Hutson, 1994). Evidence for similar regulations in fish is emerging (Loir *et al.*, 1995), and this will greatly facilitate the development of paracrine immune–endocrine models in fish.

Reproductive hormones also appear to act as immunomodulatory agents in fish. Oestradiol in fish has been shown recently to increase the susceptibility of goldfish to the haemoflagellate *Trypanosoma danilewsky* (Wang & Belosevic, 1994); previously it was established that spawning and sexually maturing fish become increasingly susceptible to a variety of infections. In mammals, a number of immunomodulatory actions of oestradiol have been demonstrated (Ito *et al.*, 1995). This has only partly been investigated in fish (Wang & Belosevic, 1994, 1995). Oestradiol in rainbow trout (Olsson *et al.*, 1995) hampers the synthesis of metallothionein, a metal-sequestering protein which also serves as an oxygen radical scavenger. This latter capacity is likely to be functionally relevant in immune–endocrine signalling, since metallothionein synthesis in mammals is regulated by factors such as IL-1, TNF-α, and is also elevated during septic shock. Finally, oestradiol affects the interrenal responsiveness of fish to stress (T.G. Pottinger & P.H.M. Balm, unpublished results), which may be one of the possible mechanisms behind the effects of maturation/spawning on immune performance and disease resistance.

Immunotoxicology is a third area of interest, and is likely to become a rich source of information about immune–endocrine interactions (Fuchs & Sanders, 1994). Several authors working with fish recognize the need to study the endocrine status of animals when investigating the mechanisms of action of toxicants on immune performance (Wester, Vethaak & van Muiswinkel, 1994). The regulation of metallothionein synthesis might also be a topic in this respect. The production of this multifunctional protein is regulated by hormones and by heavy metals. Among other functions, it detoxifies metals by distributing them to relatively safe compartments, such as the liver. Our (P.H.M. Balm & S.M.G.J. Pelgram, unpublished) results demonstrate that in tilapia (*Oreochromis mossambicus*) treated *in vivo* with IL-1 redistribution of metals such as copper, zinc and cadmium occurs. Other recent findings in this field include the induction of IL-2-like activity in carp (*Cyprinus carpio*) exposed to zinc (Ghanmi *et al.*, 1993) and the involvement of branchial granulocytes in toxic metal uptake in rainbow trout (Zia & McDonald, 1994).

Furthermore, the possibility that toxicants affect immune functions via effects on endocrine targets remains to be substantiated for fish. Many toxicants with immune effects also have endocrine effects. For instance, on the basis of their results, Lemaire-Gony, Lemaire and Pulsford (1995) discussed the operation of paracrine immune–endocrine cross-talk in the head kidneys of fish exposed to benzo(a)pyrene, which could underly the effects on immune parameters observed.

In summary, although integrated research into immune–endocrine communication in fish has been initiated relatively recently, there is compelling evidence for the existence of a functional relationship between the two systems. Since this, to a large extent, governs disease resistance, it will provide further support for stress management in fish culture.

References

Alford, P.B. III, Tomasso, J.R., Bodine, A.B. & Kendall, C. (1994). Apoptotic death of peripheral leucocytes in channel catfish: effect of confinement-induced stress. *Journal of Aquatic Animal Health*, **6**, 64–9.

Ball, J.N. & Hawkins, E.F. (1976). Adrenocortical (interrenal) responses to hypophysectomy and adenohypophysial hormones in the teleost *Poecilia latipinna*. *General and Comparative Endocrinology*, **28**, 59–70.

Balm, P.H.M. & Pottinger, T.G. (1993). Acclimation of rainbow trout (*Oncorhynchus mykiss*) to low environmental pH does not involve an activation of the pituitary–interrenal axis, but evokes adjustments in branchial ultrastructure. *Canadian Journal of Fisheries and Aquatic Sciences*, **50**, 2532–41.

Balm, P.H.M., Pepels, P., van Lieshout, E. & Wendelaar Bonga, S.E. (1993). Neuroimmunological regulation of α-MSH release in tilapia (*Oreochromis mossambicus*). *Fish Physiology and Biochemistry*, **11**, 125–30.

Balm, P.H.M., van Lieshout, E., Lokate, J. & Wendelaar Bonga, S.E. (1995a). Bacterial lipopolysaccharide (LPS) and interleukin 1 (IL-1) exert multiple physiological effects in *Oreochromis mossambicus* (Teleostei). *Journal of Comparative Physiology B*, **165**, 85–92.

Balm, P.H.M., Iger, Y., Prunet, P., Pottinger, T.G. & Wendelaar Bonga, S.E. (1995b). Skin ultrastructure in relation to prolactin and MSH function in rainbow trout (*Oncorhynchus mykiss*) exposed to environmental acidification. *Cell Tissue Research*, **279**, 351–8.

Barton, B.A. & Iwama, G.K. (1991). Physiological changes in fish from stress in aquaculture with emphasis on the response and effects of corticosteroids. *Annual Review of Fish Diseases*, **1**, 3–26.

Bayne, C.J. (1994). Adaptive thinking in immuno-nomenclature and immuno-evolution. *Immunology Today*, **15**, 598–9.

Bayne, C.J. & Levy, S. (1991). The respiratory burst of rainbow trout, *Oncorhynchus mykiss* (Walbaum), phagocytes is modulated by sympathetic neurotransmitters and the 'neuro' peptide ACTH. *Journal of Fish Biology*, **38**, 609–19.

Besedovsky, H. & Sorkin, E. (1977). Network of immune-neuroendocrine interactions. *Clinical and Experimental Immunology*, **27**, 1–12.

Betoulle, S., Troutaud, D., Khan, N., & Deschaux. (1995). Résponse anticorps, cortisolémie et prolactinémie chez la truite arc-en-ciel. *Comptes Rendus Academie de Sciences Paris*, **318**, 677–81.

Bhardwaj, R.S. & Luger, T.A. (1994). Proopiomelanocortin production by epidermal cells: evidence for an immune neuroendocrine network in the epidermis. *Archives of Dermatology Research*, **287**, 85–90.

Blalock, J.E. (1989). A molecular basis for bidirectional communication between the immune and neuroendocrine systems. *Physiological Reviews*, **69**, 1–32.

Bowley, T.J., Rance, T.A. & Baker, B.I. (1983). Measurement of immunoreactive α-melanocyte-stimulating hormone in the blood of rainbow trout kept under various conditions. *Journal of Endocrinology*, **97**, 267–75.

Byrne, P.J., Ostland, V.E., Lumsden, J.S., MacPhee, D.D. & Ferguson, H.W. (1995). Blood chemistry and acid–base balance in rainbow trout *Oncorhynchus mykiss* with experimentally-induced acute bacterial gill disease. *Fish Physiology and Biochemistry*, **14**, 509–18.

Calduch-Giner, J.A., Sitja-Bobadilla, A., Álvarez-Pellitero, P., & Pérez-Sánchez, J. (1995). Evidence for a direct action of GH on haemopoietic cells of a marine fish, the gilthead sea bream (*Sparus aurata*). *Journal of Endocrinology*, **146**, 459–67.

Catania, A. & Lipton, J.M. (1993). α-Melanocyte stimulating hormone in the modulation of host reactions. *Endocrine Reviews*, **14**, 564–76.

Cooke, H.J. (1994). Neuroimmune signalling in regulation of intestinal ion transport. *American Journal of Physiology*, **266**, G167–78.

DeKloet, E.R. & Voorhuis, T.A.M. (1992). Neuropeptides, steroid hormones, stress and reproduction. *Journal of Controlled Release*, **21**, 105–16.

Donaldson, E.M. (1981). The pituitary–interrenal axis as indicator of stress in fish. In: *Stress and Fish*, Pickering, A.D. (ed.), pp. 11–47. Academic Press, London.

Dowding, A.J. & Scholes, J. (1993). Lymphocytes and macrophages outnumber oligodendroglia in normal fish spinal cord. *Proceedings of the National Academy of Sciences, USA*, **90**, 10183–7.

Ellsaesser, C.F. & Clem, L.W. (1987). Cortisol-induced hematologic and immunologic changes in channel catfish (*Ictalurus punctatus*). *Comparative Biochemistry and Physiology A*, **87**, 405–8.

Ellsaesser, C.F. & Clem, L.W. (1994). Functionally distinct high and low molecular weight species of channel catfish and mouse IL-1. *Cytokine*, **6**, 10–20.

Faisal, M., Chiapelli, F., Ahmed, I.I., Cooper, E.L. & Weiner, H. (1989). Social confrontation 'stress' in aggresive fish is associated with an endogenous opioid-mediated suppression of proliferative response to mitogens and nonspecific cytotoxicity. *Brain, Behavior, and Immunity*, **3**, 223–33.

Faisal, M., Chiapelli, F., Ahmed, I.I., Cooper, E.L. & Weiner, H. (1992). The role of endogenous opioids in the modulation of immunosuppression in fish. *Schriftenreihe des Vereins für Wasserboden und Lufthygiene*, **89**, 785–99.

Ferguson, H.W., Morrison, D., Ostland, V.E., Lumsden, J. & Byrne, P. (1992). Responses of mucus-producing cells in gill disease of rainbow trout (*Oncorhynchus mykiss*). *Journal of Comparative Pathology*, **106**, 255–65.

Fevolden, S.E., Refstie, T. & Røed, K.H. (1992). Disease resistance in rainbow trout (*Oncorhynchus mykiss*) selected for stress response. *Aquaculture*, **104**, 19–29.

Flory, C.M. (1990). Phylogeny of neuroimmunoregulation: effects of adrenergic and cholinergic agents on the *in vitro* antibody response of the rainbow trout, *Oncorhynchus mykiss*. *Developmental and Comparative Immunology*, **14**, 283–94.

Francis, C.H. & Ellis, A.E. (1994). Production of a lymphokine (macrophage activating factor) by salmon (*Salmo salar*) leucocytes stimulated with outer membrane protein antigens of *Aeromonas salmonicida*. *Fish and Shellfish Immunology*, **4**, 489–97.

Fries, C.R. (1986). Effects of environmental stressors and immunosuppressants on immunity in *Fundulus heteroclitus*. *Amercian Zoologist*, **26**, 271–82.

Fuchs, B.A. & Sanders, V.M. (1994). The role of brain-immune interactions in immunotoxicology. *Critical Reviews in Toxicology*, **24**, 151–76.

Gaete, V., Canelo, E., Lagos, N. & Zambrano, F. (1994). Inhibitory effect of *Mycrocystis aeruginosa* toxin on ion pumps of the gill of freshwater fish. *Toxicon*, **32**, 121–7.

Gaillard, R.C. (1994). Neuroendocrine-immune system interactions. The immune-hypothalamo-pituitary-adrenal axis. *Trends in Endocrinology and Metabolism*, **5**, 303–9.

Ghanmi, Z., Rouabhia, M. & Deschaux, P. (1993). Zinc (Zn^{2+}) and fish immune response effect on carp IL2-like production and activity. *Ecotoxicology and Environmental Safety*, **25**, 236–43.

Ghoneum, M., Faisal, M., Peters, G., Ahmed, I.I. & Cooper, E.L. (1988). Suppression of natural cytotoxic cell activity by social aggressiveness in tilapia. *Developmental and Comparative Immunology*, **12**, 595–602.

Graham, S. & Secombes, C.J. (1988). The production of a macrophage-activating factor from rainbow trout *Salmo gairdneri* leucocytes. *Immunology*, **65**, 293–7.

Hamby, B.A., Huggins, E.M. Jr., Lachman, L.B., Dinarello, C.A. & Sigel, M.M. (1986). Fish lymphocytes respond to human IL-1. *Lymphokine Research*, **5**, 157–62.

Hausmann, S. (1995). 'Cytokines' in sera of lower vertebrates. *Immunology Today*, **16**, 107.

Hoffman-Goetz, L. & Pedersen, B.K. (1994). Exercise and the immune system: a model of the stress response? *Immunology Today*, **15**, 382–7.

Hooghe, R., Delhase, M., Vergani, P., Malur, A. & Hooghe-Peters, E.L. (1993). Growth hormone and prolactin are paracrine growth and differentiation factors in the haemopoietic system. *Immunology Today*, **14**, 212–14.

Hutson, J.C. (1994). Testicular macrophages. *International Review of Cytology*, **149**, 99–143

Iger, Y., Balm, P.H.M., Jenner, H.A. & Wendelaar Bonga, S.E. (1995). Cortisol induces stress-related changes in the skin of rainbow trout (*Oncorhynchus mykiss*). *General and Comparative Endocrinology*, **97**, 188–98.

Ito, I., Hayashi, T., Yamada, K., Kuzuya, M., Naito, M. & Iguchi, A. (1995). Physiological concentration of oestradiol inhibits polymorphonuclear leukocyte chemotaxis via a receptor mediated system. *Life Sciences*, **25**, 2247–53.

Jang, S.I., Hardie, L.J. & Secombes, C.J. (1994). Effects of transforming growth factor β_1 on rainbow trout *Oncorhynchus mykiss* macrophage respiratory burst activity. *Developmental and Comparative Immunology*, **18**, 315–23

Jang, S.I., Mulero, V., Hardie, L.J. & Secombes, C.J. (1995). Inhibition of rainbow trout phagocyte responsiveness to human tumor necrosis factor alpha (hTNF alpha) with monoclonal antibodies to the TNF alpha 55 kDa receptor. *Fish and Shellfish Immunology*, **5**, 61–9.

de Jong-Brink (1995). How schistosomes profit from the stress response they elicit in their hosts. *Advances in Parasitology*, **35**, 177–256.

Kaattari, S.L. (1992). Fish B lyphocytes: defining their form and function. *Annual Review of Fish Diseases*, **2**, 161–80.

Kaattari, S.L. & Tripp, R.A. (1987). Cellular mechanisms of glucocorticoid immunosuppression in salmon. *Journal of Fish Biology*, **31** (Supplement A), 129–32.

Lemaire-Gony, S., Lemaire, P. & Pulsford, A.L. (1995). Effects of cadmium and benzo(a)pyrene on the immune system, gill ATPase and EROD activity of European sea bass *Dicentrarchus labrax*. *Aquatic Toxicology*, **31**, 297–313.

Loir, M., Sourdaine, P., Mendishandagama, S.M.L.C. & Jegou, B. (1995). Cell-cell interactions in the testis of teleosts and elasmobranchs. *Microscopy Research and Technique*, **32**, 533–52.

Lotan, M. & Schwartz, M. (1994). Cross talk between the immune system and the nervous system in response to injury: implications for regeneration. *FASEB Journal*, **8**, 1026–33.

Lyons, P.D. & Blalock, J.E. (1995). The kinetics of ACTH expression in rat leukocyte subpopulations. *Journal of Neuroimmunology*, **63**, 103–12

Madden, K.S., Sanders, V.M. & Felten, D.L. (1995). Catecholamine influences and sympathetic modulation of immune responsiveness. *Annual Review of Pharmacology and Toxicology*, **35**, 417–48.

Marc, A.M., Quentel, C., Severe, A., Le Bail, P.Y. & Boeuf, G. (1995). Changes in some endocrinological and non-specific immunological parameters during seawater exposure in the brown trout. *Journal of Fish Biology*, **46**, 1065–81.

Martinez, J.L., Lopez-Doriga, M.V., Baschwitz, G.G. & Fernandez-B. De Quiros, C. (1994). Ultrastructural observations on pigmented macrophages in the epidermis of the brown trout (*Salmo trutta* L.). *Journal of Submicroscopical Cytology and Pathology*, **26**, 481–7.

Maule, A.G., Schreck, C.B. & Kaattari, S.L. (1987). Changes in the immune system of coho salmon (*Oncorhynchus kisutch*) during the parr-to-smolt transformation and after implantation of cortisol. *Canadian Journal of Fisheries and Aquatic Sciences*, **44**, 161–6.

Maule, A.G., Tripp, R.A., Kaattari, S.L. & Schreck, C.B. (1989). Stress alters immune function and disease resistance in chinook salmon (*Oncorhynchus tshawytscha*). *Journal of Endocrinology*, **120**, 135–42

Maule, A.G., Schreck, C.B. & Sharpe, C. (1993). Seasonal changes in cortisol sensitivity and glucocorticoid receptor affinity and number in leucocytes of coho salmon. *Fish Physiology and Biochemistry*, **10**, 497–506.

Maule, A.G., Schrock, R.M. & Fitzpatrick, M.S. (1994). Immune–endocrine interactions during final maturation and senescence of spring chinook salmon. *Developmental and Comparative Immunology*, **18** (Supplement 1), S69 (abstract).

McLeay, D.J. (1973). Effects of ACTH on the pituitary–interrenal axis and abundance of white blood cell types in juvenile coho salmon, *Oncorhynchus kisutch*. *General and Comparative Endocrinology*, **21**, 431–40.

Millington, G. & Buckingham, J.C. (1992). Thymic peptides and neuroendocrine–immune communication. *Journal of Endocrinology*, **133**, 163–8.

Nagae, M., Fuda, H., Ura, K., Kawamura, H., Adachi, S., Hara, A. & Yamauchi, K. (1994). The effect of cortisol administration on blood plasma immunoglobulin M (IgM) concentrations in masu salmon (*Oncorhynchus masou*). *Fish Physiology and Biochemistry*, **13**, 41–8.

Nakano, T., Sato, M. & Takeuchi, M. (1995). Unique molecular properties of superoxide dismutase from teleost fish skin. *FEBS Letters*, **360**, 197–201.

Narnaware, Y.K., Baker, B.I. & Tomlinson, M.G. (1994). The effect of various stresses, corticosteroids and adrenergic reagents on phagocytosis in the rainbow trout *Oncorhynchus mykiss*. *Fish Physiology and Biochemistry*, **13**, 31–40.

Ndoye, A., Troutaud, D., Rougier, F. & Deschaux, P. (1991). Neuroimmunology in fish. *Advances in Neuroimmunology*, **1**, 242–51.

Olsson, P.-E., Kling, P., Petterson, C. & Silversand, C. (1995). Interaction of cadmium and oestradiol-17β on metallothionein and vitellogenin synthesis in rainbow trout (*Oncorhynchus mykiss*). *Biochemical Journal*, **307**, 197–203.

Oswald, I.P., Eltoum, I., Wynn, T.A., Schwartz, B., Caspar, P., Paulin, D., Sher, A. & James, S.L. (1994). Endothelial cells are activated by cytokine treatment to kill an intravascular parasite, *Schistosoma mansoni*, through the production of nitric oxide. *Proceedings of the National Academy of Sciences, USA*, **91**, 999–1003.

Ottaviani, E., Franchini, A. & Fontanili, P. (1994). The effect of corticotropin-releasing factor and pro-opiomelanocortin-derived peptides on the phagocytosis of molluscan hemocytes. *Experientia*, **50**, 857–9.

Ottaviani, E., Caselgrandi, E. & Franceschi, C. (1995a). Cytokines and evolution: *in vitro* effects of IL-1α, IL-1β, TNF-α and TNF-β on an ancestral type of stress response. *Biochemical and Biophysical Research Communications*, **207**, 288–92.

Ottaviani, E., Capriglione, T. & Franceschi, C. (1995b). Invertebrate and vertebrate immune cells express pro-opiomelanocortin (POMC) mRNA. *Brain, Behavior, and Immunity*, **9**, 1–8.

Peters, G., Nüssgen, A., Raabe, A. & Möck, A. (1991). Social stress induces structural and functional alterations of phagocytes in rainbow trout (*Oncorhynchus mykiss*). *Fish and Shellfish Immunology*, **1**, 17–31.

Pickering, A.D. (1984). Cortisol-induced lymphocytopenia in brown trout, *Salmo trutta* L. *General and Comparative Endocrinology*, **53**, 252–9.

Pickering, A.D. (1989). Environmental stress and the survival of brown trout, *Salmo trutta*. *Freshwater Biology*, **21**, 47–55.

Pickford, G.E., Srivastava, A.K., Slicher, A.M. & Pang, P.K.T. (1971). The stress response in the abundance of circulating leucocytes in the killifish, *Fundulus heteroclitus*. III. The role of adrenal cortex and a concluding discussion of the leucocyte-stress syndrome. *Journal of Experimental Zoology*, **177**, 109–18.

Plytycz, B. & Seljelid, R. (1995). Nonself as a stressor – inflammation as a stress reaction. *Immunology Today*, **16**, 110–11.

Pohajdak, B., Dixon, B. & Stuart, G.R. (1993). Immune system. In *Biochemistry and Molecular Biology of Fishes*, Volume 2. Hochachka, P.W. & Mommsen, T.P. (eds.) pp. 191–205. Elsevier Science Publishers, Amsterdam.

Rand, T.G. & Cone, D.K. (1990). Effects of *Ichthyophonus hoferi* on condition indices and blood chemistry of experimentally infected rainbow trout (*Oncorhynchus mykiss*). *Journal of Wildlife Diseases*, **26**, 323–8.

Rasquin, P. (1951). Effects of carp pituitary and mammalian ACTH on the endocrine and lymphoid systems of the teleost *Astyanax mexicanus*. *Journal of Experimental Zoology*, **117**, 317–57.

Rivier, C. (1995). Influence of immune signals on the hypothalamic-pituitary axis of the rodent. *Frontiers in Neuroendocrinology*, **16**, 151–82.

Rothwell, N.J. & Hopkins, S.J. (1995). Cytokines and the nervous system II: actions and mechanisms of action. *Trends in Neurosciences*, **18**, 130–6.

Rowley, A.F., Hunt, T.C., Page, M. & Mainwaring, G. (1988). Fish. In *Vertebrate Blood Cells* Rowley, A.F. & Ratcliff, N.A. (eds.) pp. 257–334. Cambridge University Press, Cambridge.

Sakai, M., Kobayashi, M. & Kawauchi, H. (1995). *In vitro* activation of rainbow trout, *Oncorhynchus mykiss*, phagocytic cells by growth hormone, prolactin and somatolactin. In *Proceedings of the The Nordic Symposium on Fish Immunology 1995*. p. 66 (abstract). May 24–7, 1995, Reykjavik, Iceland.

Scharrer, B. (1991). Neuroimmunology: the importance and role of a comparative approach. *Advances in Neuroimmunology*, **1**, 1–6.

Schreck, C.B. & Bradford, S. (1990). Interrenal corticosteroid production: potential regulation by the immune system in the salmonid. *Progress in Clinical and Biological Research*, **342**, 480–6.

Secombes, C.J. (1994). Enhancement of fish phagocyte activity. *Fish and Shellfish Immunology*, **4**, 421–36.

Secombes, C.J. & Fletcher, T.C. (1992). The role of phagocytes in the protective mechanisms of fish. *Annual Review of Fish Diseases*, **2**, 53–71.

Secombes, C.J., Fletcher, T.C., White, A., Costello, M.J., Stagg, R. & Houlihan, D.F. (1992). Effects of sewage sludge on immune responses in the dab, *Limanda limanda* (L.). *Aquatic Toxicology*, **23**, 217–30.

Sharp, B. & Linner, K. (1993). What do we know about the expression of proopiomelanocortin transcripts and related peptides in lymphoid tissue? *Endocrinology*, **133**, 1921A–1921B.

Silver, R., Silverman, A.J., Vitkovic, L. & Lederhandler, I.I. (1996). Mast cells in the brain: evidence and functional significance. *Trends in Neurosciences*, **19**, 25–31.

Sim, G.K. (1995). Intraepithelial lymphocytes and the immune system. *Advances in Immunology*, **58**, 297–343.

Slater, C.H. & Schreck, C.B. (1993). Testosterone alters the immune response of chinook salmon, *Oncorhynchus tshawytscha*. *General and Comparative Endocrinology*, **89**, 291–8.

Slater, C.H., Fitzpatrick, M.S. & Schreck, C.B. (1995). Characterization of an androgen receptor in salmonid lymphocytes: possible link to androgen-induced immunosuppression. *General and Comparative Endocrinology*, **100**, 218–25.

Snieszko, S.F. (1974). The effects of environmental stress on outbreaks of infectious diseases of fishes. *Journal of Fish Biology*, **6**, 197–208.

Stave, J.W. & Robertson, B.S. (1985). Hydrocortisone suppresses the chemiluminescent response of striped bass phagocytes. *Developmental and Comparative Immunology*, **9**, 77–84

Sternberg, E.M., Scott Young, W. III, Bernardini, R., Calogero, A.E., Chrousos, G.P., Gold, P.W. & Wilder, R.L. (1989). A central nervous sytem defect in biosynthesis of corticotropin-releasing hormone is associated with susceptibility to streptococcal cell wall-induced arthritis in Lewis rats. *Proceedings of the National Academy of Sciences, USA.*, **86**, 4771–5

Stickney, R.R. (1986). Tilapia. In *Culture of nonsalmonid Freshwater Fishes*. Stickney, R.R. (ed.) pp. 58–72. CRC Press, Boca Raton, USA.

Sumner, F.B. & Douderoff, P. (1938). The effects of light and dark backgrounds upon the incidence of a seemingly infectious disease in fishes. *Proceedings of the National Academy of Science, USA*, **24**, 463–6.

Sundar, S.K., Cierpial, M.A., Kamaraju, L.S., Long, S., Hsieh, S., Lorenz, C., Aaron, M., Ritchie, J.C. & Weiss, J.M. (1991). Human immunodeficiency virus glycoprotein (gp120) infused into rat brain induces interleukin 1 to elevate pituitary–adrenal activity and decreases peripheral cellular immune responses. *Proceedings of the National Academy of Science, USA*, **88**, 11246–50.

Thompson, I., White, A., Fletcher, T.C., Houlihan, D.F. & Secombes, C.J. (1993). The effect of stress on the immune response of Atlantic salmon (*Salmo salar* L.) fed different diets containing different amounts of vitamin C. *Aquaculture*, **114**, 1–18.

Tilders, F.J.H., de Rijk, R.H., van Dam, A.-M., Vincent V.A.M., Schotanus, K. & Persoons, J.H.A. (1994). Activation of the hypothalamus–pituitary–adrenal axis by bacterial endotoxins: routes and intermediate signals. *Psychoneuroendocrinology*, **19**, 209–32.

Tripp, R.A., Maule, A.G., Schreck, C.B. & Kaattari, S.L. (1987). Cortisol mediated suppression of salmonid lymphocyte response *in vitro*. *Developmental and Comparative Immunology*, **11**, 565–76.

Vindrola, O., Mayer, A.M.S., Citera, G., Spitzer, J.A. & Espinoza, L.R. (1994). Prohormone convertases PC2 and PC3 in rat neutrophils and macrophages. *Neuropeptides*, **27**, 235–44

Walters, G.R. & Plumb, J.A. (1980). Environmental stress and bacterial infection in channel catfish, *Ictalurus punctatus* Rafinesque. *Journal of Fish Biology*, **17**, 177–85.

Wang, R. & Belosevic, M. (1994). Oestradiol increases susceptibility of goldfish to *Trypanosoma danilewskyi*. *Developmental and Comparative Immunology*, **18**, 377–87.

Wang, R. & Belosevic, M. (1995). The *in vitro* effects of oestradiol and cortisol on the function of a long-term goldfish macrophage cell line. *Developmental and Comparative Immunology*, **19**, 327–36.

Wedemeyer, G. (1969). Pituitary activation by bacterial endotoxins in the rainbow trout (*Salmo gairdneri*). *Journal of Bacteriology*, **100**, 542–3.

Wedemeyer, G. (1970). The role of stress in the disease resistance of fishes. *Special Publication American Fisheries Society*, **5**, 30–5.

Wester, P.W., Vethaak, A.D. & van Muiswinkel, W.B. (1994). Fish as biomarkers in immunotoxicology. *Toxicology*, **86**, 213–32.

White, A. & Fletcher, T.C. (1985). The influence of hormones and inflammatory agents on C-reactive protein, cortisol, and alanine aminotransferase in the plaice (*Pleuronectes platessa* L.). *Comparative Biochemistry and Physiology C*, **80**, 99–104.

Wilder, R.L. (1995). Neuroendocrine-immune system interactions and autoimmunity. *Annual Review of Immunology*, **13**, 307–38.

Zaccone, G., Fasulo, S. & Ainis, L. (1995). Neuroendocrine epithelial cell system in respiratory organs of air-breathing and teleost fishes. *International Review of Cytology*, **157**, 277–314.

Zia, S. & McDonald, D.G. (1994). Role of gills and gill chloride cells in metal uptake in the freshwater-adapted rainbow trout, *Oncorhynchus mykiss*. *Canadian Journal of Fisheries and Aquatic Sciences*, **51**, 2482–92.

T.C. FLETCHER

Dietary effects on stress and health

Introduction

Responsiveness to stress is evident in fish from an early age. Rainbow trout (*Oncorhynchus mykiss*) at 10 °C are 6 weeks old when they first show a post-stress increase in cortisol levels. Hatching occurred 4 weeks after fertilization, which would indicate that the hypothalamic–pituitary–interrenal axis is responsive to stress 2 weeks after hatching and 1 week before the start of exogenous feeding (Barry *et al.*, 1995). The elevation of plasma levels of corticosteroids, mainly cortisol in teleost fish, is considered a primary response to stress, as is the secretion of catecholamines (Barton & Iwama, 1991). Their release is the start of a sequence of responses which are a necessary part of survival. Low levels of stress can be dealt with by avoidance, acclimation and compensation (Chiappelli *et al.*, 1993) but sometimes the challenge to the maintenance of homeostasis is so great that the limits of tolerance are exceeded and pathological processes ensue. Moberg (1992) discussed this biological cost of stress which can ultimately lead to reproductive loss, a reduction in growth, abnormal behaviour and disease.

It is generally accepted that elevated levels of cortisol can be immuno-suppressive, resulting in increased vulnerability to pathogens (Barton & Iwama, 1991). Immune function and resistance to the fish pathogen *Vibrio anguillarum* are depressed in juvenile chinook salmon (*Oncorhynchus tshawytscha*) within 4 h of exposure to acute stress, when plasma cortisol levels are at their highest (Maule *et al.*, 1989), although at 24 h disease resistance was enhanced and cortisol levels were similar to controls. These experiments illustrate the variability in the response of the immune system to stress and, as Maule *et al.* (1989) point out, although cortisol may have a direct immunosuppress-ive effect, there may be other hormonally driven events which are participating in the outcome. An interesting finding has been that carp (*Cyprinus carpio*) exhibiting protective immunity have similar mortality rates to their non-immunized counterparts, if they are injected with a corticosteroid prior to challenge with a protozoan parasite (Houghton & Matthews, 1986). When Nagae *et al.* (1994) fed diets containing cortisol to masu salmon (*Oncorhynchus masou*) the plasma concentration of antibody-associated immunoglobulin M was specifically suppressed although α1-protein and total protein were not affected. The chapter by P.H.M. Balm, this volume, provides further details of the interactions between endocrine and immune systems in fish.

It is with the rapid growth and expansion of aquaculture that a parallel increased interest in stress has became more evident. Intensive culture systems frequently bring with them a range of stressors that fish would not normally encounter in the wild and the option of movement away from a hostile environment is not generally available to them. It is in the commercial interests of the fish farmer to try to alleviate the stress, the effects of which become apparent in pathological conditions. This can be achieved by careful husbandry but it is not possible to eliminate entirely some associated stressors and optimal conditions for culture of many species of fish have yet to be established.

A requirement of intensive aquaculture is that the fish must be fed. If, with this one operation, the health of the fish could be sufficiently enhanced to withstand the pathologies to which they seem prone, then an important advance would have been made. It is therefore not surprising that there is a growing body of research directed towards this goal. Concurrently, the conditions for the successful oral administration of vaccines as feed supplements are also being sought, primarily to eliminate the stress of handling, which is inherent in immersion and injection techniques. Although this provides a more focused research objective than the complex interactions between nutrition and immunity, the methods of oral immunization have, as yet, proved largely ineffective (Bøgwald *et al.*, 1994). The potential of dietary manipulation to enhance disease resistance and immune function is still mainly at an exploratory stage although the significance of adequate nutrition is well established and formulations are constantly reviewed to provide optimal concentrations of the essential components for maintenance diets (National Research Council, 1993). Nutritional diseases are often difficult to recognize because superimposed infectious states, which are clinically more obvious, can mask the underlying nutritional basis. It is rare in practice for a deficiency of only one nutrient to occur but it is only by experimental studies of single-deficiency conditions for individual species that the basis of nutritional pathologies can be recognized. Roberts and Bullock (1989) describe the pathology of dietary deficiency and imbalance in macronutrients, i.e. proteins, carbohydrate and lipids, and in micronutrients, i.e. vitamins and minerals.

The existing publications dealing with immunity and diet have been reviewed comprehensively by Landolt (1989), Blazer (1992) and Lall and Olivier (1993), while Waagbø (1994) deals specifically with the Atlantic salmon, *Salmo salar*. The relationship between stress and diet and immunity is a topic that has been discussed only briefly in these reviews. This is in part due to the paucity of reported experiments in this area. Those that do exist and are described here are heavily weighted towards certain nutrients, with vitamin C pre-eminent, but it is hoped that, with time, the range of dietary constituents tested will be extended.

Macronutrients

The basis for any dietary alleviation of the stress response must, ultimately, involve some effect of the component at the hormonal and/ or neural level of control. Although there is speculation in this area, as yet no experiments appear to have been designed to elucidate the underlying molecular mechanisms and the literature is largely descriptive of what happens when various nutrients are manipulated, often in an empirical manner. This applies to the macronutrients, where dietary concentrations have been linked with stress and disease.

Carbohydrates

Carbohydrate metabolism is significantly affected as a secondary stress response (Mazeaud, Mazeaud & Donaldson, 1977) and this fact might be expected to attract some attention to the manipulation of this dietary component. Hemre, Lambertsen and Lie (1991) have reported on the effect of dietary carbohydrate on the response to stress. They worked with cod (*Gadus morhua*), which has a low ability to utilize carbohydrate in the feed. Using extruded wheat as the carbohydrate source, 440-g cod were fed diets either lacking carbohydrate or containing 25% carbohydrate on a dry weight basis. After 9 weeks on the diets, the fish were netted and transported for 2 h in oxygenated water and then sampled over the following 96 h. Perhaps the most relevant result was that both groups showed similar cortisol responses with maximal levels of approximately 14.5 ng ml^{-1} plasma within 30 min. There was, however, a significant difference in plasma glucose, with the concentrations in resting fish reflecting the dietary levels. Although levels in both groups reached a maximum between 1 and 3 h after stress, in the carbohydrate-deficient group there was a rise from a resting concentration of 3.3 to 4.9 mmol glucose l^{-1} plasma, while in the carbohydrate-fed cod values rose from 4.4 to 10.3 and were still at 7.1 mmol glucose l^{-1} plasma after 96 h, at which time the carbohydrate-deficient group was back to 3.9 mmol l^{-1}. There was also a significant increase in glycogen in white muscle and livers from the carbohydrate-fed fish. The effects of stress on carbohydrate metabolism were therefore influenced by the previous dietary carbohydrate history of the cod.

If dietary carbohydrate causes any malfunctioning in the regulation of the stress response then there might be a subsequent effect on disease resistance. Although physical stressors were not involved, Waagbø *et al.* (1994) have looked at the effects of dietary carbohydrate on immunity and resistance to disease challenge in the Atlantic salmon. They used a range of diets, containing from 0% to 30% (dry weight) of highly available carbohydrate and reported two experiments; firstly with adult (0.5 kg) and then with juvenile (3 g) salmon. Serum glucose concentration was not significantly related to dietary carbohydrate in the adults but there was a positive correlation in the juveniles. Serum cortisol in the juveniles was not measured but in the adults it increased linearly (4.9 to 10.1 ng ml^{-1} serum) when they were fed from 5 to 30% carbohydrate respectively. Fish fed on the carbohydrate-deficient diet had serum cortisol values of 9.4 ng ml^{-1}, similar to those of fish fed 30% carbohydrate; these results are comparable to those measured from cod fed diets either deficient in, or containing 25% carbohydrate (Hemre *et al.*, 1991). The antibody response of adult fish immunized with *Vibrio salmonicida* was not affected by the carbohydrate concentration and there was no correlation with the cortisol levels. Juvenile fish challenged with different concentrations of *V. anguillarum* in the water showed no statistical differences in mortality. In the adults challenged by intraperitoneal injection of *Aeromonas salmonicida*, mortality ranged from 32 to 44%, with fish on the 10% carbohydrate diet showing the lowest mortality. Waagbø *et al.* (1994) concluded that dietary levels of carbohydrate do not significantly affect disease resistance in the Atlantic salmon, although qualifying this with the need for further research using fish in a more active growth phase and different strains at different seasons.

Certain carbohydrates, although not normally dietary ingredients, can act as immunopotentiators, enhancing non-specific disease resistance in fish. Their effects on the neuroendocrine system do not yet appear to have been investigated, although an immunoactive synthetic peptide has been reported to restore the phagocytic activity of cells from rainbow trout immunosuppressed with cortisol (Kitao & Yoshida, 1986).

Particular attention has been given to β-glucans, which are branched β-1,3- and β-1,6-linked glucose polymers occurring in the cell walls of most yeast and fungi (Robertsen, Engstad & Jørgensen, 1994). Evidence shows that they stimulate the non-specific defence mechanisms of fish, increasing resistance against bacterial infections (Robertsen *et al.*, 1994). Based on the work of Ainsworth *et al.* (1991), in which phagocytes were found to be more resistant than lymphocytes to temperature stress, Chen and Ainsworth (1992) reasoned that compounds potentiating non-specific defence mechanisms would have considerable prophylactic value in stressful aquaculture conditions. Published results mainly describe administration of glucans by injection but there are now good indications that their incorporation into feeds will also enhance disease resistance (Raa *et al.*, 1992). However, when Ainsworth, Mao and Boyle (1994) fed a β-glucan from *Schizophyllum commune* to channel catfish, they were not able to enhance resistance to the pathogen *Edwardsiella ictaluri*, although there was a higher mortality rate among the controls during the first 2–3 days when deaths started to occur. There were also significantly higher antibody titres among fish fed the β-glucan at 0.1% supplementation of the diet. Other carbohydrates being tested as immunostimulants include chitin (Sakai *et al.*, 1992) and chitosan (Anderson & Siwicki, 1994).

Proteins

A reduction or cessation of growth with weight loss is a consequence of inadequate dietary protein, with the withdrawal of protein from less vital tissues to maintain the functioning of those more important (Wilson, 1989). Two months of starvation did not reduce the concentration of catecholamines in the chromaffin tissue of the rainbow trout (Reid, Furimsky & Perry, 1994) so that existing protein stores must have provided the tyrosine and any other nutrients required for biosynthesis in the absence of a dietary supply. Many constituents of the immune system, such as immunoglobulins, cytokines and enzymes, are proteins, so that a deficiency might be expected to impair the immune response.

Antibody titres were, however, unrelated to dietary protein in experiments conducted by Kiron *et al.* (1993a). They fed adult rainbow trout isocaloric diets either lacking protein or containing 20, 35 and 50% protein, for 3 months prior to immunization with *A. salmonicida*. As there was no reduction in antibody in the group fed a protein-deficient diet, Kiron *et al.* (1993a) concluded that immunoglobulin synthesis was preferentially maintained although total protein in the blood reflected dietary levels. Kiron et al. (1993a) used the same protein diets as described above to rear fry for 3 weeks before exposure to infectious haematopoietic necrosis virus. Although no statistics were presented, slightly fewer mortalities occurred with 20% and 35% protein diets than with 50% protein diets, but most mortalities occurred in the protein-deficient group. Kiron et al. (1993a) compared their results with those of Hardy, Halver and Brannon (1979), who found more mortalities in chinook salmon fingerlings fed diets with 65% protein than with 30% protein, when challenged with V. anguillarum. In this case the results were complicated by the fact that the two protein diets also contained pyridoxine (vitamin B_6), at four concentrations. When the vitamin was present at 40 mg kg^{-1} in the high-protein diet, then the fewest mortalities occurred. Such results would argue for the need for balance in all the constituents of a diet. When Albrektsen et al. (1995) fed pre-smolt Atlantic salmon a 70% fishmeal diet supplemented with different amounts of pyridoxine (0–160 mg kg^{-1} diet) for 5 months and challenged with A. salmonicida, there was no clear effect from feeding the different amounts. Kiron et al. (1993b) reported another experiment where rainbow trout were fed diets lacking protein or containing 35% and 50% protein. After 3 months on the diets, fish leucocytes were tested for natural-killer activity but this protective mechanism was not affected by dietary protein.

Dietary protein levels have been reported to affect the severity of cryptobiosis, a protozoan infection, in rainbow trout (Li & Woo, 1991). Parasitaemia was significantly higher in fish fed a 52% protein diet than in those on 37 or 22% protein diets. Plasma proteins were also significantly higher in the group fed the 52% protein diet and Li and Woo concluded that higher levels of plasma proteins increased the multiplication of the parasite so that disease was more severe in fish on the highest protein diet: 65% mortality compared with 25% and 30% mortalities in fish on the 37% and 22% protein diets respectively. Anorexia was a consequence of infection and Li and Woo (1991) hypothesized that, by reducing the intake of dietary protein, the multiplication rate of the parasite was reduced. Thus, considerations of the effects of the diet on the parasite should be superimposed on those of the effects of the diet on the host. Beisel (1987) listed mechanisms by which malnutrition may cause an infection to be (or appear to be) less severe and these included a reduction in the availability of specific nutrients or host metabolites that are essential for the replication of microorganisms. Infection only followed challenge of rainbow trout with the bacterial gill disease organism Flavobacterium branchiophilum when the fish were fed following exposure (MacPhee *et al.*, 1995). Cumulative mortalities of up to 63% were measured, compared with 0–2% in fish unfed from the day of exposure. The authors proposed a physiological basis for their observations, rather than a deterioration in water quality. The consumption of food would result in the excretion of carbon dioxide, nitrogenous waste and other metabolic products which could accumulate in the unstirred layer overlying the gill lamellar epithelium and so create conditions favouring the attachment and multiplication of the bacteria.

Beisel (1987), although describing mammalian systems, pointed out that with starvation in the absence of disease, metabolic adaptation is rapid and mechanisms are employed by the body to conserve protein nitrogen to a maximum degree. The body is therefore able to maintain the competence of the immune system, often for long periods. This was shown when African catfish (*Clarias gariepinus*), starved or fed daily at maintenance levels of 1.7 g kg^{-1} metabolic weight (kg$^{0.8}$), could still mount an antibody response when immunized with *Yersinia ruckeri* O-antigen after 87 days on these treatments. There was no significant difference between the agglutinating titres of these two groups (log$_2$ titres 2.49 and 2.02 respectively, 1 week post-immunization) but fish fed at 15.1 g kg^{-1} metabolic weight showed peak antibody titres of 6.19 after 1 week (Henken, Tigchelaar & Van Muiswinkel, 1987). These authors did not introduce challenge organisms so it was not known whether resistance was significantly reduced in the fish deprived of food or fed at maintenance levels. In aquaculture, when food intake is depressed, it is usually as a result of handling, low temperatures or low oxygen and at such times the fish are more susceptible to disease.

Feed allowance is designed for maximal growth rates and it has not yet been established whether these also provide optimal immunocompetence. Although a group of fish are fed at a specific level, not all the fish receive an equal share of the ration, this being determined by the feeding hierarchy. Manipulation of the level of feeding can affect this situation: results with rainbow trout (McCarthy, Carter & Houlihan, 1992) suggest that the strength of the feeding hierarchy and the variability in individual consumption decreases as food availability increases. In a dominance hierarchy, subordinate fish have been found to be more susceptible to pathogen invasion compared with the dominant individuals, which are less stressed (Ejike & Schreck, 1980; Chiappelli *et al.*, 1993).

Rather than adjust the feeding level or amounts of dietary protein, the approach taken by Pouliot and de la Noüe (1989) was to feed protein of either animal or vegetable origin in an attempt to ameliorate the stress of hypoxia in rainbow trout. In mammals the biosynthesis of the neurotransmitters serotonin and catecholamines in the central nervous system is influenced by the bioavailability of their amino acid precursors, tryptophan and tyrosine respectively. Some resistance to stress is apparent if the organism maintains the initial equilibrium between serotonin and noradrenaline in the central nervous system. This balance will be dependent on the plasma supply of their precursor amino acids which, in turn, depends on the amino acid profile of the diet. Pouliot and de la Noüe (1989) found that rainbow trout that were adapted to diets of the same nitrogen and energy contents for 149 days and then subjected to hypoxic conditions maintained their pre-stress equilibrium between the serotonin- and catecholamine-dependent pathways if fed the animal protein. Their feed intake also remained normal or returned to normal after stress, whereas that of fish fed vegetable protein did not. Similar experiments were reported by Neji *et al.* (1993), with the addition of infection with *A. salmonicida* to hypoxic stress. As before, fish subjected to hypoxia only maintained a normal feed intake if fed animal protein, the availability of tryptophan being a possible factor. Feed intake of all groups was reduced by infection but, regardless of oxygen availability, mortalities were always significantly higher among fish fed vegetable protein. The relationship between dietary tryptophan and stress responses in fish seems worthy of further study. Serotonin-dependent activity was found to be elevated in the brains of stressed Arctic char (*Salvelinus alpinus*) but starvation had no effect on the utilization of this transmitter system (Winberg, Nilsson & Olsén, 1992).

Lipids

Dietary lipids provide the essential fatty acids necessary for normal growth and development. Fish cannot synthesize either linolenic $[18 : 3(n - 3)]$ or linoleic acid $[18 : 2(n - 6)]$ and hence one or both of these fatty acids must be provided by the diet, depending on the essential fatty acid requirements (discussed in National Research Council, 1993). The polyunsaturated fatty acids (PUFA) from fish tissues are predominantly of the $n - 3$ series, of which, quantitatively, the main components are eicosapentaenoic acid $[20 : 5(n - 3)]$ and docosahexaenoic acid $[22 : 6(n - 3)]$. The diet must supply either the PUFA themselves or their metabolic precursors. Certain fatty acids are essential components of membrane phospholipids; a high PUFA content conferring fluidity to membranes at low temperatures. This appears to be best provided by the highly unsaturated $n - 3$ series fatty acids. There are much smaller quantities of PUFA of the $n - 6$

series in fish but arachidonic acid $[20 : 4(n - 6)]$ is important in the generation of eicosanoids, which are involved in immunity.

An essential fatty acid-deficiency symptom reported by Castell *et al.* (1972) in relation to rainbow trout held on diets low in $n - 3$ fatty acids is a shock syndrome that is exhibited after handling or tank stress. Symptoms appeared after 1–3 months on deficient diets and were severe in fish supplemented with 1% linoleic acid but were completely prevented by diets containing 0.5% or more of linolenic acid. Bell *et al.* (1991) reported a transportation-induced shock syndrome that caused 30% mortality in Atlantic salmon fed for 16 weeks on a fishmeal-based diet where lipid was provided by sunflower oil but no deaths occurred when fish oil was used. Although both diets contained $n - 3$ fatty acids, they only represented 9.9% of the sunflower oil but 23.8% of the fish oil. In comparison, $n - 6$ fatty acids were present at 43.6% and 2.5% respectively. It appears that a diet with a low $(n - 3)/(n - 6)$ ratio (0.2, sunflower oil; 9.4 fish oil) causes changes in fatty acid metabolism which are damaging to salmonids, especially when they are stressed. Severe heart lesions were also observed to occur with the sunflower oil diet. Although the eicosanoid status of the salmon was not reported, Bell *et al.* (1991) speculated that this would be affected by the relative amounts of the PUFA present. Both arachidonic $(n - 6)$ and eicosapentaenoic $(n - 3)$ acid act as precursors for eicosanoid generation but give rise to different series of leukotrienes, lipoxins and prostanoids with differing biological potencies (Ashton *et al.*, 1994). These authors found that changes could occur in fatty acid composition and the eicosanoid-generating capacity of leucocytes from rainbow trout, within 4–8 weeks of changing the lipid composition of their diets.

The type of dietary fatty acid appeared to affect the survival of larval mahimahi, *Coryphaena hippurus*, after the stress of being held out of the water for 1 or 2 min (Kraul *et al.*, 1993). The larvae were fed copepods or enriched *Artemia* but recovery from stress depended on a high content of docosahexaenoic acid $(n - 3)$ in this diet: high levels of eicosapentaenoic acid did not confer stress resistance if the level of docosahexaenoic acid was low.

The possibility that dietary levels of $n - 3$ PUFA might influence responses to some of the physiological stresses common in aquaculture has been investigated by McKenzie *et al.* (1994). Diets were enriched (15%) with either fish (menhaden) oil (FOD) as a source of $n - 3$ fatty acids or with saturated fatty acids, such as hydrogenated coconut oil (COD), and fed to the Adriatic sturgeon (*Acipenser naccarii*). Hypoxia can occur in aquaculture and the response of the sturgeon to such conditions was measured (Fig. 1). Fish fed FOD consumed less oxygen under normal conditions than those on the COD diet or even a commercial diet. This situation was maintained during hypoxia, when aerobic metabolism was unchanged for the FOD group but reduced in

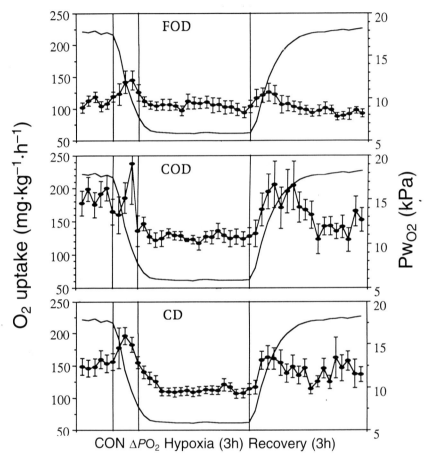

Fig. 1. Mean (±SEM) O_2 uptake in Adriatic sturgeon (*Acipenser naccarii*) fed a control diet (CD), a diet enriched in fish (menhaden) oil (FOD) or a diet enriched in coconut oil (COD), in normoxia, hypoxia (6.6 kPa O_2 tension), and recovery; $n = 7$ in all cases. Closed symbols, O_2 uptake; solid line, water Pw_{O_2}; CON, control; $\triangle PO_2$, period when water Pw_{O_2} was declining. Reproduced from McKenzie *et al.* (1994), with permission from the authors and from the Fish Physiology Association.

the groups not receiving high levels of $n - 3$ supplementation. The addition of similar fats to the diet of tilapia (*Oreochromis nilotica*) did not affect resting oxygen uptake but $n - 3$-supplemented diets led to a significant reduction in ammonia excretion following exercise. These results therefore gave some encouragement to the possibility of dietary amelioration of the response to some stressors.

The role of lipids in fish immunity is not well understood but they can modulate cellular responses by influencing the physical properties of membranes and hence the membrane-associated enzymes and receptor sites. The biosynthesis of the immunoactive eicosanoids will also be determined by the lipids available. These factors all contribute to disease resistance but results from different laboratories are often conflicting. A major cause is probably the variation in the type and composition of the dietary lipids and even their susceptibility to oxidation.

Sheldon and Blazer (1991) found that enhanced bactericidal activity of channel catfish macrophages was correlated with increasing levels of dietary $n - 3$ fatty acids. The natural-killer-cell-like activity of rainbow trout leucocytes was also enhanced (almost trebled) when the fish were fed a diet containing $n - 3$ fatty acids compared with totally saturated palmitic acid (Kiron *et al.*, 1993*b*).

Different results emerge from other *in vivo* experiments where $n - 3$ fatty acids often appear to exert an immunosuppressive action, as found by Erdal *et al.* (1991) in Atlantic salmon fed increasing amounts of $n - 3$ fatty acids from menhaden oil or fish oil concentrate. Fish immunized with *Yersinia ruckeri* gave significantly lower antibody titres when fed on $n - 3$-enriched diets and had a lower likelihood of survival when challenged with the organism. A high $n - 3$ fatty acid content did not have any protective effect when fish were challenged with *Vibrio salmonicida* although Erdal *et al.* (1991) quote other workers finding reduced mortality when Atlantic salmon fed increased amounts of $n - 3$ fatty acids were challenged with cold water vibriosis. Erdal *et al.* (1991) concluded that their working temperature of 10–12 °C was perhaps too high for these potential protective effects. Li *et al.* (1994) found reduced resistance to *Edwardsiella ictaluri* when channel catfish were fed increased amounts of $n - 3$, from menhaden oil at 24 °C, although antibody titres arising from the challenge were not affected. Fracalossi and Lovell (1994) have also examined survival of channel catfish fed different dietary lipids to challenge with *E. ictaluri* but this time at both 28 °C and 17 °C. Although antibody titres were highest with a menhaden oil diet (28 °C fish) they found significantly reduced resistance to infection in fish on the highest $n - 3$ fatty acid diets (linseed and menhaden oils) at the higher temperature but no significant difference at 17 °C, suggesting that metabolism of fatty acids for immunoregulatory functions is temperature related. A possible explanation, advanced by Fracalossi and Lovell (1994) for the reduced resistance at warmer temperatures, was a difference in eicosanoid synthesis with competitive inhibition of $n - 6$ metabolism by the $n - 3$ fatty acids. The balance between different fatty acids and the temperature at which experiments are conducted may well be the key to the correct interpretation.

Micronutrients

These essential components have been found to be implicated in many aspects of the immune system of mammals (Bendich & Chandra, 1990) but similar studies have not been made with fish although there is a growing body of literature on vitamins. Most of the fish studies have been of the effects of single nutrients, although Hilton (1989) reviewed nutrient interactions in fish, considering vitamin–vitamin, vitamin–mineral, mineral–mineral and micronutrient–macronutrient interactions. This useful information covered both dietary and metabolic aspects but not immunological effects.

Minerals

In his comprehensive review of minerals in fish, Lall (1989) pointed out that it is sometimes difficult to demonstrate a mineral deficiency because fish have the ability to absorb some inorganic elements (calcium, magnesium, sodium, potassium, iron, zinc, copper and selenium) not only from their diet but also from their environment, both in fresh water and sea water. Aquaculture procedures are often associated with stressors causing osmoregulatory disturbances which upset the water and mineral balance of the fish (Mazeaud et al., 1977; see also chapter by G. McDonald and L. Milligan, this volume). A dietary approach to the alleviation of the stress of salinity change has occasionally been taken and examples are given by Lall (1989). More recently, Pelletier and Besner (1992) found improved survival rates of brook charr (*Salvelinus fontinalis*) fed diets containing 8% or 12% NaCl, before transfer to sea water.

The macroelements calcium, magnesium, sodium, potassium and phosphorus (with chlorine) are required in gram quantities by the body but the trace elements are normally only present at milligram or microgram quantities per kilogram body mass. Inorganic elements are involved in normal life processes and are components of hormones and enzymes (National Research Council, 1993) but their exact role in immune functions is unknown. There are few reports linking minerals with immunity in fish and, apart from in disease, their dietary manipulation does not appear to have been tested with stressors associated with aquaculture. The selection of individual minerals for immunological study appears to have been mainly arbitrary and results too few to be conclusive.

There was evidence from Lall *et al.* (1985) that the prevalence of bacterial kidney disease was reduced in Atlantic salmon smolts fed diets containing increased levels of iodine and fluorine (both at 4.5 mg kg^{-1} diet, compared with 1.5 mg kg^{-1}) but their absorption and bioavailability were affected by mineral concentrations in the diet and water. Kiron *et al.* (1993*b*) examined the effect of dietary zinc in their studies of natural-killer cell activity in rainbow trout. The mean cytotoxicity value for cells from a zinc-deficient group was 5.07% compared with 16.96% for fish fed zinc at 40 mg kg^{-1} diet. The role of zinc was also examined in channel catfish (Scarpa, Gatlin & Lewis, 1992); it was found that feeding a diet deficient in zinc (2.0 mg zinc kg^{-1} diet) for 10 weeks resulted in no mortalities following challenge with an intraperitoneal injection of *Aeromonas hydrophila*. Perhaps this reflected a nutritional deficiency for the pathogen as there were 41.7% mortalities in the replete diet (20.0 mg zinc kg^{-1}). The same challenge to fish on a calcium-deficient diet (0.02% of diet) resulted in 50% mortality when the fish were in water low in calcium, but only 16.7% when calcium represented 2.5% of the diet.

Iron is of particular interest because the ability of some microorganisms to establish an infection depends on the availability of this element in their host (Lall & Olivier, 1993). A significant increase in the virulence of *V. anguillarum* was demonstrated in eels injected with ferric ammonium citrate (Nakai *et al.*, 1987). The amount of free iron in fish is, presumably, limited as it is maintained by iron-binding proteins (transferrin, lactoferrin) but bacteria are able to synthesize iron chelators, siderophores, which compete with transferrin for the iron. Ravndal *et al.* (1994) found higher serum iron concentrations in Atlantic salmon susceptible to infection by *Vibrio* species. Little is known of the effect of iron on immunity in fish but control of dietary iron could influence disease resistance. Ravndal *et al.* (1994) quote findings of Rørvik and colleagues (1992), who found a significant positive correlation between a natural outbreak of furunculosis and concentrations of dietary iron but not serum iron, suggesting that availability of iron in the gastrointestinal tract may be important for multiplication of *A. salmonicida*. Iron deficiency causes microcytic anaemia in several fish species but is not common in culture conditions where diets contain adequate iron. The concentration of iron in common feedstuffs is highly variable and can be elevated by contamination from ferrous metal during processing (Lall, 1989). The level of iron in commercial diets is a matter of importance if, by reduction, disease resistance can be enhanced.

Another aspect of dietary iron that could be involved in pollution stressors is the demonstration of the selective regulatory effects of iron on the induction of different cytochrome P450 isozymes in Atlantic salmon (Goksøyr, Bjørnevik & Maage, 1994). These isozymes can act on a range of xenobiotics, and the possibility that iron loading may unfavourably affect their regulation led Goksøyr et al. (1994) to speculate that their findings might be related to reported detrimental effects of high levels of dietary iron on fish health. Vitamin C (ascorbic acid: AA) is involved in the metabolism of iron in fish (Hilton, 1989) and Maage et al. (1990) observed that vitamin C-deficient rainbow trout were anaemic, despite elevated iron concentrations in the liver.

Vitamins

Vitamins have been popular candidates for dietary manipulation in the quest for enhanced immunocompetence and improved disease resistance, and none more so than vitamin C. Blazer (1992) reviewed much of the research done in this area and Sandnes (1991) also discussed immunology and disease resistance when reviewing vitamin C in fish nutrition. Vitamin C is also the one vitamin that has been promoted as having a positive role in the amelioration of stress.

A number of factors have linked vitamin C and stress in fish over the years. Corticosteroids are associated with the anterior kidney where adrenal function is located in the interrenal tissues under the control of pituitary adrenocorticotropic hormone (ACTH). The anterior kidney is also rich in AA, although reflecting dietary levels (White et al., 1993). Kidney AA levels in coho salmon were found to decrease from 89 $\mu g\ g^{-1}$ kidney tissue in controls to 48 $\mu g\ g^{-1}$ following 15 min of forced exertion and this response could be mimicked by the injection of ACTH (Wedemeyer, 1969). Mild stressing over 2 h caused a decrease in kidney AA over the first 20 min followed by a return to almost the initial value by the end of the 2-h period. As there was no concomitant increase in plasma levels of AA, Wedemeyer (1969) suggested that AA could be used in steroid biosynthesis, especially as serum cortisol levels increased as AA decreased. Kitabchi (1967) had hypothesized that high levels of AA had an inhibitory role in steroid synthesis, by preventing the conversion of unsaturated fatty acids into cholesterol esters which are incorporated into steroids. It was therefore a logical step to suggest that increasing the AA loading of the fish might prevent the severity of a stress response. Wedemeyer (1969) reported that raising kidney concentrations (by injection) to as much as 1 mg g^{-1} did not prevent AA depletion when the fish were stressed.

There are reports in the literature which would support the idea of increasing AA concentrations in fish under stress. Ishibashi *et al.* (1992) fed Japanese parrot fish, *Oplegnathus fasciatus*, diets supplemented with 0, 75 and 300 mg AA 100 g^{-1} for 2 weeks before subjecting half of each dietary group to hypoxic conditions every 3 or 4 days for 16 weeks. Tolerance to hypoxia was evaluated by a behavioural response of 'lying down' and this had reached almost 100% within 12 weeks in the AA-deficient group compared with less than 20% in the other two groups. In the AA-deficient group, associated deficiency symptoms appeared earlier and were more severe in the stressed than in the non-stressed fish and, even on the 75 mg AA diet, stressed fish showed inferior growth compared to their controls, with a significant reduction in plasma AA and a marked reduction in the concentration in the kidneys and gills. Fish on the high 300 mg AA diet were not significantly affected by the stressor and Ishibashi *et al.* (1992) concluded that dietary AA ameliorated the effects of hypoxic stress in the species studied.

Mazik, Brandt and Tomasso (1987) fed channel catfish fry diets containing 0, 78 or 390 mg vitamin C kg^{-1} diet for 120 days and stressed them with increased ammonia concentrations, hypoxia and net confinement. The fish on diets with AA exhibited similar tolerance to the first two stressors while that of the deficient group was lower, but no mortalities occurred in any of the groups subjected to netting stress. As there was no difference in the responses to the stressors between the two supplemented groups, the role of AA in the alleviation of stress might have been better demonstrated if supplementation at a higher level had also been included: Ishibashi *et al.* (1992) were using 3 g AA kg^{-1} diet.

Other workers have not been able to relate dietary concentrations of AA to the stress response. Gouillou-Coustans and Guillaume (1993) found that stress conditions did not exacerbate deficiency symptoms in turbot (*Scophthalmus maximus*). Liver AA stores were not depleted faster in stressed AA-deficient fish, as might be expected if AA was implicated in hormone synthesis and Gouillou-Coustans and Guillaume (1993) concluded that the consequences of stress do not clearly depend on the AA status of the turbot.

Serum cortisol levels increased in Atlantic salmon following physical stress but were not significantly influenced by the AA status of the fish, which had been fed for 4 weeks on diets lacking AA or supplemented with calcium ascorbate-2-monophosphate, equivalent to 500 mg AA kg^{-1} diet (Sandnes & Waagbø, 1991). The concentrations of AA in liver and kidney reflected their dietary intake but there was no significant change in these concentrations following stress. Thompson *et al.* (1993) did not measure cortisol in Atlantic salmon fed at different dietary levels of AA and subjected to a 2-h confinement stress but glucose was used as a stress indicator. There was no significant difference in the glucose levels between fish on the different diets or in the magnitude of their hyperglycaemic response to stress. Liver AA levels were not affected by the stress but various immunological functions were affected. The leucocyte respiratory burst and bactericidal activity were depressed, while plasma bactericidal activity was enhanced, by stress but all were unaffected by vitamin C status. Dabrowski and Ciereszko (1993) found that in rainbow trout subjected to hauling stress, liver and kidney AA levels increased whereas levels of the less biologically active dehydroascorbate decreased in the liver, suggesting some interorgan transport.

At present there do not seem to be any biochemical studies which confirm an involvement of AA in the biosynthesis of corticosteroids or catecholamines in fish. Work with guinea-pigs has led Laney, Levy and Kipp (1990) to conclude that the absolute level of AA in the adrenal glands is not critical for steroidogenesis. Results from Dabrowska *et al.* (1991) indicate that, in the carp, post-stress cortisol levels are also independent of the vitamin C status of the fish. In mammals, AA has been implicated in the synthesis of neurotransmitters. In bovine tissue it has been demonstrated that AA is required for the biosynthesis of noradrenaline from dopamine (Levine *et al.*, 1985). In rainbow trout there appears to be a significant correlation between brain AA status and brain serotonin levels after 12 weeks on diets containing between 0 and 320 mg AA kg^{-1} diet, although this relationship did not hold at 24 weeks (Johnston, MacDonald & Hilton, 1989). Even if AA has not been proved to alleviate the response of fish to certain types of stressors, there seems little doubt that increased dietary levels do contribute to disease resistance and enhance certain immunological responses in fish (Blazer, 1992; Waagbø, 1994).

Vitamins A, B, and E have been reviewed by Blazer (1992) in relation to their role in disease resistance but they do not appear to have been tested as alleviators of stress in fish. Thompson *et al.* (1994) fed Atlantic salmon diets containing vitamin A at 0.37, 1.95 and

15 mg kg^{-1} diet for 4 months. No significant difference was found in the resistance to *A. salmonicida* although mortalities were highest among fish on the 0.37 mg vitamin diet. This might be a reflection of the lack of effect of dietary vitamin A on a number of immunological factors, including phagocyte respiratory burst and bactericidal activity, eicosanoid, lymphokine and antibody production, as well as serum complement and lysozyme activity. Leucocyte migration and serum bactericidal activity were significantly reduced in fish fed low levels of vitamin A. There was no evidence to suggest that diets rich in vitamin A would enhance disease resistance (Thompson *et al.*, 1994). As already described (p. 228), Albrektsen *et al.* (1995) found that increased concentrations of dietary vitamin B$_6$ did not enhance disease resistance in Atlantic salmon.

Vitamin E (α-tocopherol) is the major lipid-soluble membrane antioxidant of cells, in which it is considered to protect unsaturated bonds of fatty acids from oxidation. Fish tissues have high concentrations of unsaturated fatty acids which are vulnerable to lipid peroxidation, and Hilton (1989) discusses the relationship between PUFA and vitamin E. There is also synergism between this vitamin and selenium and it also appears to interact with vitamin C (Hilton, 1989). It is perhaps because of these relationships that there is often variance in the results from different groups describing the protective effects of vitamin E (Lall & Olivier, 1993). In his review of salmonids, Waagbø (1994) concluded that vitamin-E-deficient diets resulted in immunological malfunction and reduced disease resistance. Obach, Quentel and Baudin Laurencin (1993) fed sea bass (*Dicentrarchus labrax*) with diets containing different levels of vitamin E and fresh or oxidized oil and found that disease resistance was not affected by dietary treatment. There are reports of vitamin E reducing stress in mammals (Tengerdy, 1989) and Watson and Petro (1982) found significantly reduced serum corticosteroid concentrations in mice fed vitamin E (4 g kg^{-1} diet) but similar experiments with fish do not appear to have been reported.

Conclusions

The objective of much of the work reviewed in this chapter was to enhance disease resistance through dietary manipulation. Overall, the evidence supports the use of some dietary components in this role but their most suitable level of feeding has still to be established when the dietary requirements for optimal growth do not always coincide with those for optimal functioning of the immune system. Management of stress is another aspect of improving fish survival but from the results presented it would be difficult to formulate a diet that could be used with success for the variety of fish now cultured. Progress would be made if more was known of the action of various dietary components at the molecular level and also of nutrient interactions, some of which have been described by Hilton (1989). It is probable, however, that the use of dietary manipulation will be superseded by breeding fish, both for attenuated stress responsiveness (discussed by T.G. Pottinger and A.D. Pickering, this volume) and disease resistance. The techniques of biotechnology can also make possible the transfer of genes encoding for proteins associated with disease resistance. These approaches will provide the next advances for fish health but there is still a place for nutritional studies. The diet could provide the vehicle for the introduction of engineered vaccines and adjuvants, while the nutrient composition might be designed to complement different stages of a vaccination programme.

References

Ainsworth, A.J., Dexiang, C., Waterstrat, P.R. & Greenway, T. (1991). Effect of temperature on the immune system of channel catfish (*Ictalurus punctatus*). I. Leucocyte distribution and phagocyte function in the anterior kidney at 10 °C. *Comparative Biochemistry and Physiology*, **100A**, 907–12.

Ainsworth, A.J., Mao, C.P. & Boyle, C.R. (1994). Immune response enhancement in channel catfish, *Ictalurus punctatus*, using β-glucan from *Schizophyllum commune*. In *Modulators of Fish Immune Responses; Volume 1*. Stolen, J.S. & Fletcher, T.C. (eds.) pp. 67–81. SOS Publications, Fair Haven, New Jersey.

Albrektsen, S., Sandnes, K., Glette, J. & Waagbø, R. (1995). Influence of dietary vitamin B_6 on tissue vitamin B_6 contents and immunity in Atlantic salmon, *Salmo salar* L. *Aquaculture Research*, **26**, 331–9.

Anderson, D.P. & Siwicki, A.K. (1994). Duration of protection against *Aeromonas salmonicida* in brook trout immunostimulated with glucan or chitosan by injection or immersion. *Progressive Fish-Culturist*, **56**, 258–61.

Ashton, I., Clements, K., Barrow, S.E., Secombes, C.J. & Rowley, A.F. (1994). Effects of dietary fatty acids on eicosanoid-generating capacity, fatty acid composition and chemotactic activity of rainbow trout (*Oncorhynchus mykiss*) leucocytes. *Biochimica et Biophysica Acta*, **1214**, 253–62.

Barry, T.P., Malison, J.A., Held, J.A. & Parrish, J.J. (1995). Ontogeny of the cortisol stress response in larval rainbow trout. *General and Comparative Endocrinology*, **97**, 57–65.

Barton, B.A. & Iwama, G.K. (1991). Physiological changes in fish from stress in aquaculture with emphasis on the response and effects of corticosteroids. *Annual Review of Fish Diseases*, **1**, 3–26.

Beisel, W.R. (1987). Nutritional and metabolic factors in host responses to infection. In *Immunopharmacology of Infectious Diseases: Vaccine Adjuvants and Modulators of Non-Specific Resistance*. Majde, J.A. (ed.) pp. 51–60. Alan R. Liss, New York.

Bell, J.G., McVicar, A.H., Park, M.T. & Sargent, J.R. (1991). High dietary linoleic acid affects the fatty acid compositions of individual phospholipids from tissues of Atlantic salmon (*Salmo salar*): association with stress susceptibility and cardiac lesion. *Journal of Nutrition*, **121**, 1163–72.

Bendich, A. & Chandra, R.K. (eds.) (1990). Micronutrients and immune functions. *Annals of the New York Academy of Sciences*, **587**, 1–320.

Blazer, V.S. (1992). Nutrition and disease resistance in fish. *Annual Review of Fish Diseases*, **2**, 309–23.

Bøgwald, J., Stensvåg, K., Stuge, T.B. & Jørgensen, T.Ø. (1994). Tissue localisation and immune responses in Atlantic salmon, *Salmo salar* L., after oral administration of *Aeromonas salmonicida*, *Vibrio anguillarum* and *Vibrio salmonicida* antigens. *Fish and Shellfish Immunology*, **4**, 353–68.

Castell, J.D., Sinnhuber, R.O., Wales, J.H. & Lee, D.J. (1972). Essential fatty acids in the diet of rainbow trout (*Salmo gairdneri*): growth, feed conversion and some gross deficiency symptoms. *Journal of Nutrition*, **102**, 77–86.

Chen, D. & Ainsworth, A.J. (1992). Glucan administration potentiates immune defence mechanisms of channel catfish, *Ictalurus punctatus* Rafinesque. *Journal of Fish Diseases*, **15**, 295–304.

Chiappelli, F., Franceschi, C., Ottaviani, E., Farnè, M. & Faisal, M. (1993). Phylogeny of the neuroendocrine–immune system: fish and shellfish as model systems for social interaction stress research in humans. *Annual Review of Fish Diseases*, **3**, 327–46.

Dabrowska, H., Dabrowski, K., Meyer-Burgdorff, K., Hanke, W. & Gunther, K.-D. (1991). The effect of large doses of vitamin C and magnesium on stress responses in common carp, *Cyprinus carpio*. *Comparative Biochemistry and Physiology*, **99A**, 681–5.

Dabrowski, K. & Ciereszko, A. (1993). Influence of fish size, origin, and stress on ascorbate concentration in vital tissues of hatchery rainbow trout. *Progressive Fish-Culturist*, **55**, 109–13.

Ejike, C. & Schreck, C.B. (1980). Stress and social hierarchy rank in coho salmon. *Transactions of the American Fisheries Society*, **109**, 423–6.

Erdal, J.I., Evensen, Ø., Kaurstad, O.K., Lillehaug, A., Solbakken, R. & Thorud, K. (1991). Relationship between diet and immune response in Atlantic salmon (*Salmo salar* L.) after feeding various levels of ascorbic acid and omega-3 fatty acids. *Aquaculture*, **98**, 363–79.

Fracalossi, D.M. & Lovell, R.T. (1994). Dietary lipid sources influence responses of channel catfish (*Ictalurus punctatus*) to challenge with the pathogen *Edwardsiella ictaluri*. *Aquaculture*, **119**, 287–98.

Goksøyr, A., Bjørnevik, M. & Maage, A. (1994). Effects of dietary iron concentrations on the cytochrome P450 system of Atlantic salmon (*Salmo salar*). *Canadian Journal of Fisheries and Aquatic Sciences*, **51**, 315–20.

Gouillou-Coustans, M.F. & Guillaume, J. (1993). Effect of a non specific stressor on the symptoms of ascorbic acid deficiency in turbot (*Scophthalmus maximus*). In *Fish nutrition in practice (Proceedings of IVth International Symposium on Fish Nutrition and Feeding, June 1991, Biarritz)*. Kaushik, S.J. & Luquet, P. (eds.) pp. 209–13. INRA, Paris.

Hardy, R.W., Halver, J.E. & Brannon, E.L. (1979). Effect of dietary protein level on the pyridoxine requirement and disease resistance of chinook salmon. In *Finfish Nutrition and Fish Feed Technology*, Volume 1. Halver, J.E. & Tiews, K. (eds.) pp. 253–60. Heenemann, Berlin.

Hemre, G.I., Lambertsen, G. & Lie, Ø. (1991). The effect of dietary carbohydrate on the stress response in cod (*Gadus morhua*). *Aquaculture*, **95**, 319–28.

Henken, A.M., Tigchelaar, A.J. & Van Muiswinkel, W.B. (1987). Effects of feeding level on antibody production in African catfish, *Clarias gariepinus* Burchell, after injection of *Yersinia ruckeri* O-antigen. *Journal of Fish Diseases*, **11**, 85–8.

Hilton, J.W. (1989). The interaction of vitamins, minerals and diet composition in the diet of fish. *Aquaculture*, **79**, 223–44.

Houghton, G. & Matthews, R.A. (1986). Immunosuppression of carp (*Cyprinus carpio* L.) to Ichthyophthiriasis using the corticosteroid triamcinolone acetonide. *Veterinary Immunology and Immunopathology*, **12**, 413–19.

Ishibashi, Y., Kato, K., Ikeda, S., Murata, O., Nasu, T. & Kumai, H. (1992). Effects of dietary ascorbic acid on tolerance to intermittent hypoxic stress in Japanese parrot fish. *Nippon Suisan Gakkaishi*, **58**, 2147–52.

Johnston, W.L., MacDonald, E. & Hilton, J.W. (1989). Relationships between dietary ascorbic acid status and deficiency, weight gain and brain neurotransmitter levels in juvenile rainbow trout, *Salmo gairdneri*. *Fish Physiology and Biochemistry*, **6**, 353–65.

Kiron, V., Fukuda, H., Takeuchi, T. & Watanabe, T. (1993*a*). Dietary protein related humoral immune response and disease resistance of rainbow trout, *Oncorhynchus mykiss*. In *Fish Nutrition in Practice*

(Proceedings of IVth International Symposium on Fish Nutrition and Feeding, June 1991, Biarritz). Kaushik, S.J. & Luquet, P. (eds.), pp. 24–7. INRA, Paris.

Kiron, V., Gunji, A., Okamoto, N., Satoh, S., Ikeda, Y. & Watanabe, T. (1993*b*). Dietary nutrient dependent variations on natural-killer activity of the leucocytes of rainbow trout. *Gyobyo Kenkyu*, **28**, 71–6.

Kitabchi, A.E. (1967). Ascorbic acid in steroidogenesis. *Nature*, **215**, 1385–6.

Kitao, T. & Yoshida, Y. (1986). Effect of an immunopotentiator on *Aeromonas salmonicida* infection in rainbow trout (*Salmo gairdneri*). *Veterinary Immunology and Immunopathology*, **12**, 287–96.

Kraul, S., Brittain, K., Cantrell, R., Nagao, T., Ako, H., Ogasawara, A. & Kitagawa, H. (1993). Nutritional factors affecting stress resistance in the larval mahimahi *Coryphaena hippurus*. *Journal of the World Aquaculture Society*, **24**, 186–93.

Lall, S.P. (1989). The minerals. In *Fish Nutrition*. Halver, J.E. (ed.) pp. 219–57. Academic Press, San Diego.

Lall, S.P. & Olivier, G. (1993). Role of micronutrients in immune response and disease resistance in fish. In *Fish Nutrition in Practice (Proceedings of IVth International Symposium on Fish Nutrition and Feeding, June 1991, Biarritz).* Kaushik, S.J. & Luquet, P. (eds.) pp. 101–18. INRA, Paris.

Lall, S.P., Paterson, W.D., Hines, J.A. & Adams, N.J. (1985). Control of bacterial kidney disease in Atlantic salmon, *Salmo salar* L., by dietary modification. *Journal of Fish Diseases*, **8**, 113–24.

Landolt, M.L. (1989). The relationship between diet and the immune response of fish. *Aquaculture*, **79**, 193–206.

Laney, P.H., Levy, J.A. & Kipp, D.E. (1990). Plasma cortisol and adrenal ascorbic acid levels after ACTH treatment with a high intake of ascorbic acid in the guinea pig. *Annals of Nutrition and Metabolism*, **34**, 85–92.

Levine, M., Morita, K., Heldman, E. & Pollard, H.B. (1985). Ascorbic acid regulation of norepinephrine biosynthesis in isolated chromaffin granules from bovine adrenal medulla. *Journal of Biological Chemistry*, **260**, 15598–603.

Li, M.H., Wise, D.J., Johnson, M.R. & Robinson, E.H. (1994). Dietary menhaden oil reduced resistance of channel catfish (*Ictalurus punctatus*) to *Edwardsiella ictaluri*. *Aquaculture*, **128**, 335–44.

Li, S. & Woo, T.K. (1991). Anorexia reduces the severity of cryptobiosis in *Oncorhynchus mykiss*. *Journal of Parasitology*, **77**, 467–71.

Maage, A., Waagbø, R., Olsson, P.E., Julshamn, K. & Sandnes, K. (1990). Ascorbate-2-sulfate as a dietary vitamin C source for Atlantic salmon (*Salmo salar*): 2. Effects of dietary levels and immunization on the metabolism of trace elements. *Fish Physiology and Biochemistry*, **8**, 429–36.

MacPhee, D.D., Ostland, V.E. Lumsden, J.S., Derksen, J. & Ferguson, H.W. (1995). Influence of feeding on the development of bacterial gill disease in rainbow trout *Oncorhynchus mykiss*. *Diseases of Aquatic Organisms*, **21**, 163–70.

Maule, A.G., Tripp, R.A., Kaattari, S.L. & Schreck, C.B. (1989). Stress alters immune function and disease resistance in chinook salmon *(Oncorhynchus tshawytscha)*. *Journal of Endocrinology*, **120**, 135–42.

Mazeaud, M. M., Mazeaud, F. & Donaldson, E.M. (1977). Primary and secondary effects of stress in fish: some new data with a general review. *Transactions of the American Fisheries Society*, **106**, 201–12.

Mazik, P.M., Brandt, T.M. & Tomasso, J.R. (1987). Effects of dietary vitamin C on growth, caudal fin development, and tolerance of aquaculture-related stressors in channel catfish. *Progressive Fish-Culturist*, **49**, 13–16.

McCarthy, I.D., Carter, C.G. & Houlihan, D.F. (1992). The effect of feeding hierarchy on individual variability in daily feeding of rainbow trout, *Oncorhynchus mykiss* (Walbaum). *Journal of Fish Biology*, **41**, 257–63.

McKenzie, D.J., Piraccini, G., Taylor, E.W., Steffensen, J.F., Bronzi, P. & Bolis, L. (1994). Effect of dietary lipids on responses to stress in fish. In *High Performance Fish (Proceedings of an International Fish Physiology Symposium, July 1994, University of British Columbia)*. MacKinlay, D.D. (ed.) pp. 431–36. Fish Physiology Association, Vancouver.

Moberg, G.P. (1992). Stress induced pathologies in fish: the cost of stress. *NOAA Technical Report NMFS*, **111**, 131–4.

Nagae, M., Fuda, H., Ura, K., Kawamura, H., Adachi, S., Hara, A. & Yamauchi, K. (1994). The effect of cortisol administration on blood plasma immunoglobulin M (IgM) concentrations in masu salmon *(Oncorhynchus masou)*. *Fish Physiology and Biochemistry*, **13**, 41–8.

Nakai, T., Kanno, T., Cruz, E.R. & Muroga, K. (1987). The effects of iron compounds on the virulence of *Vibrio anguillarum* in Japanese eels and ayu. *Fish Pathology*, **22**, 185–9.

National Research Council (1993). *Nutrient Requirements of Fish*. National Academy Press, Washington DC.

Neji, H., Naimi, N., Lallier, R. & de la Noüe, J. (1993). Relationships between feeding, hypoxia, digestibility and experimentally induced furunculosis in rainbow trout. In *Fish Nutrition in Practice (Proceedings of IVth International Symposium on Fish Nutrition and Feeding, June 1991, Biarritz)*. Kaushik, S.J. & Luquet, P. (eds.) pp. 187–97. INRA, Paris.

Obach, A., Quentel, C. & Baudin Laurencin, F. (1993). Effects of alpha-tocopherol and dietary oxidized fish oil on the immune response of sea bass *Dicentrarchus labrax*. *Diseases of Aquatic Organisms*, **15**, 175–85.

Pelletier, D. & Besner, M. (1992). The effect of salty diets and gradual transfer to sea water on osmotic adaptation, gill Na^+,K^+-ATPase activation, and survival of brook trout, *Salvelinus fontinalis,* Mitchill. *Journal of Fish Biology*, **41**, 791–803.

Pouliot, T. & de la Noüe, J. (1989). Feed intake, digestibility and brain neurotransmitters of rainbow trout under hypoxia. *Aquaculture*, **79**, 317–27.

Raa, J., Roerstad, G., Engstad, R. & Robertsen, B. (1992). The use of immunostimulants to increase resistance of aquatic organisms to microbial infections. In *Diseases in Asian Aquaculture, I.* Shariff, M., Subasinghe, R.P. & Arthur, J.R. (eds.), pp. 39–50. Fish Health Section, Asian Fisheries Society, Manila.

Ravndal, J., Løvold, T., Bentsen, H.B., Røed, K.H., Gjedrem, T. & Rørvik, K-A. (1994). Serum iron levels in farmed Atlantic salmon: family variation and associations with disease resistance. *Aquaculture*, **125**, 37–45.

Reid, S.G., Furimsky, M. & Perry, S.F. (1994). The effects of repeated physical stress or fasting on catecholamine storage and release in the rainbow trout, *Oncorhynchus mykiss*. *Journal of Fish Biology*, **45**, 365–78.

Roberts, R.S. & Bullock, A.M. (1989). Nutritional pathology. In *Fish Nutrition,* Halver, J.E. (ed.) pp. 423–73. Academic Press, San Diego.

Robertsen, B., Engstad, R.E. & Jørgensen, J.B. (1994). β-Glucans as immunostimulants in fish. In *Modulators of Fish Immune Responses*, Volume 1. Stolen, J.S. & Fletcher, T.C. (eds.) pp. 83–99. SOS Publications, Fair Haven, New Jersey.

Sakai, M., Kamiya, H., Ishii, S., Atsuta, S. & Kobayashi, M. (1992). The immunostimulating effects of chitin in rainbow trout, *Oncorhynchus mykiss*. In *Diseases in Asian Aquaculture, I.* Shariff, M., Subasinghe, R.P. & Arthur, J.R. (eds.) pp. 413–17. Fish Health Section, Asian Fisheries Society, Manila.

Sandnes, K. (1991). Vitamin C in fish nutrition – a review. *Fiskeridirektoratets Skrifter, Serie Ernœring*, **4**, 3–32.

Sandnes, K. & Waagbø, R. (1991). Effects of dietary vitamin C and physical stress on head kidney and liver ascorbic acid, serum cortisol, glucose and haematology in Atlantic salmon *(Salmo salar)*. *Fiskeridirektoratets Skrifter, Serie Ernœring*, **4**, 41–9.

Scarpa, J., Gatlin, D.M. & Lewis, D.H. (1992). Effects of dietary zinc and calcium on select immune functions of channel catfish. *Journal of Aquatic Animal Health*, **4**, 24–31.

Sheldon, W.M. & Blazer, V.S. (1991). Influence of dietary lipid and temperature on bactericidal activity of channel catfish macrophages. *Journal of Aquatic Animal Health*, **3**, 87–93.

Tengerdy, R.P. (1989). Vitamin E, immune response, and disease resistance. *Annals of the New York Academy of Sciences*, **570**, 335–44.

Thompson, I., White, A., Fletcher, T.C., Houlihan, D.F. & Secombes, C.J. (1993).The effect of stress on the immune response of the Atlantic salmon *(Salmo salar* L.*)* fed diets containing different amounts of vitamin C. *Aquaculture*, **114**, 1–18.

Thompson, I., Fletcher, T.C., Houlihan, D.F. & Secombes, C.J. (1994). The effect of dietary vitamin A on the immunocompetence of Atlantic salmon (*Salmo salar* L.). *Fish Physiology and Biochemistry*, **12**, 513–23.

Waagbø, R. (1994).The impact of nutritional factors on the immune system in Atlantic salmon, *Salmo salar* L.: a review. *Aquaculture and Fisheries Management*, **25**, 175–97.

Waagbø, R., Glette, J., Sandnes, K. & Hemre, G.I. (1994). Influence of dietary carbohydrate on blood chemistry, immunity and disease resistance in Atlantic salmon, *Salmo salar* L. *Journal of Fish Diseases*, **17**, 245–58.

Watson, R.R. & Petro, T.M. (1982). Cellular immune respones, corticosteroid levels, and resistance to *Listeria monocytogenes* and murine leukemia in mice fed a high vitamin E diet. *Annals of the New York Academy of Sciences*, **393**, 205–8.

Wedemeyer, G. (1969). Stress-induced ascorbic acid depletion and cortisol production in two salmonid fishes. *Comparative Biochemistry and Physiology*, **29**, 1247–51.

White, A., Fletcher, T.C., Secombes, C.J. & Houlihan, D.F. (1993). The effect of different dietary levels of vitamins C and E on their tissue levels in the Atlantic salmon, *Salmo salar* L. In *Fish nutrition in Practice (Proceedings of IVth International Symposium on Fish Nutrition and Feeding, June 1991, Biarritz)*. Kaushik, S.J. & Luquet, P. (eds.) pp. 203–7. INRA, Paris.

Wilson, R.P. (1989). Amino acids and proteins. In *Fish Nutrition*. Halver, J.E. (ed.) pp. 111–51. Academic Press, San Diego.

Winberg, S., Nilsson, G.E. & Olsén, K.H. (1992). The effect of stress and starvation on brain serotonin utilization in Arctic charr *(Salvelinus alpinus)*. *Journal of Experimental Biology*, **165**, 229–39.

J.D. MORGAN and G.K. IWAMA

Measurements of stressed states in the field

Introduction

There are a number of methods available for evaluating the effects of stress on fish (Adams, 1990a; Wedemeyer, Barton & McLeay, 1990), but many of these techniques are only appropriate for research or clinical laboratories, because they involve relatively sophisticated procedures and require expensive equipment to be performed. The problem of assessing stressed states in fish outside the laboratory has been a challenge for both fish culturists and fisheries biologists. The purpose of this chapter is to outline a number of simple, reliable and portable methods to detect stress and assess the general condition of fish under field conditions.[1] The 'field' in this case is defined as fish hatcheries and saltwater rearing sites, where power is usually available and bench space may or may not be available, as well as natural settings, such as rivers and lakes, where neither sources of electric power or laboratory facilities are likely to occur. The portable instruments described here have been validated against standard laboratory-based assays and field tested at aquaculture facilities and a detailed description of these studies can be found elsewhere (Iwama, Morgan & Barton, 1995).

Overview and rationale of features measured

The simple field methods described in this chapter include a number of physiological, hematological and physical condition indicators for monitoring variables associated with stress in fish. A routine monitoring program that incorporates a suite of these variables is recommended when studying a particular fish population, rather than relying on a single indicator which reflects only one level of biological organization (Adams, 1990b).

---

[1]Trade and company names mentioned in this chapter are for information purposes only and do not imply endorsement by the authors.

Glucose

Blood glucose is probably the most commonly measured secondary change that occurs during the stress response in fish (Wedemeyer *et al.*, 1990). Blood glucose increases as the stress response begins in order to provide energy to the animal for the 'fight-or-flight' reaction. Elevated blood glucose levels are initiated and sustained by the actions of epinephrine and cortisol, respectively, on the liver and muscle. These stress hormones, however, are expensive to measure and, in the case of cortisol, the techniques usually involve the use of radioactive tracers that require special licensing and facilities to perform. Glucose, on the other hand, is relatively easy and inexpensive to determine, and analysis can be performed using portable meters developed for human diabetic patients.

Hematology

Hematological features can also be useful as indicators of stress in fish. Changes in blood hemoglobin, red blood cell numbers or hematocrit (% RBC packed volume) values following acute stress may indicate that hemodilution or hemoconcentration has occurred due to impaired osmoregulation. Increases in these variables following stress may also be due to increased red blood cell numbers resulting from splenic contraction, or erythrocyte swelling through the actions of epinephrine, to facilitate gas transfer (Nikinmaa, 1990; see also chapter by G. McDonald and L. Milligan, this volume). Decreases may further indicate anemia, or a reduction in circulating red blood cell numbers, resulting from infectious diseases (Cardwell & Smith, 1971) .

The abundance of circulating white blood cells, responsible for antibody production, provides a good indication of the health of the fish. Decreases in leukocyte (white blood cell) numbers commonly occur during the stress response and can be determined precisely by differential cell counts from microscopic examination of fixed blood smears (Houston, 1990). A less accurate, but more rapid, method of estimating white blood cell numbers is to determine the leukocrit, or packed cell volume of white blood cells in a microhematocrit tube (McLeay & Gordon, 1977). The chapter by P.H.M. Balm, this volume, examines the links between environmental stress, the endocrine system and the immune system.

Plasma constituents

Effects of stress on ionic regulation are normally determined by measuring plasma chloride and sodium concentrations with a chloridometer, flame photometer, atomic absorption spectrophotometer, or ion chromatograph. Of these instruments, the chloridometer is the most portable and easiest to use. Low plasma chloride levels (hypochloremia) often result from stress and can become life threatening if they fall below 90 mEq l^{-1} (Wedemeyer *et al.*, 1990). Ionic and osmotic impacts of stress are discussed in more detail in the chapter by G. McDonald and L. Milligan, this volume. Changes in plasma protein levels can be used as an indicator of impaired water balance, as well as nutritional imbalance due to chronic stress (Wedemeyer & Yasutake, 1977), and can be determined easily using a portable refractometer.

Physical condition indices

There are a number of physical condition indicators that provide information regarding the general health and stress status of fish populations, and they can be easily measured in the field. These include condition factor, growth and various organosomatic indices such as ratios of liver, spleen, viscera and gonad mass to total body mass (Goede & Barton, 1990). Decreases in these ratios may indicate chronic stress, although some of the indices will vary naturally with season.

Goede and Barton (1990) described an empirical, autopsy-based assessment procedure as an indicator of the health and condition of fish. This health–condition index system can be used to monitor the physical characteristics of the fins, eyes, gills, pseudobranchs, thymus, liver, kidney, spleen, gut, and bile, as well as the extent of fat deposition. Various states of organ appearance and texture are categorized using a numerical ranking procedure. A health–condition profile (HCP) is established by averaging the scores for a representative sample of the population. A normal appearance of vital organs is assumed to indicate that a fish population is coping well with its environment. Long-term exposure to a stressor may result in physiological alterations which will be expressed at the organ level. The autopsy procedure is rapid, simple to perform, requires little training and does not require costly, sophisticated equipment. It has been used successfully in fish hatcheries and in feral trout populations in Utah (Goede, 1989), and in studies of juvenile chinook salmon (*Oncorhynchus tshawytscha*) in the Columbia River (Novotny & Beeman, 1990). Recently, the autopsy system was modified to a health assessment index (HAI) system which provides a quantitative index so that statistical comparisons can be made between data sets. This approach has been used to evaluate the general health status of fish populations in the Tennessee River Valley (Adams, Brown & Goede, 1993).

Description of simple field methods

Glucose

Minilab

The Ames Minilab (Miles Canada Inc., Etobicoke, Ontario) is a fixed-wavelength (546 nm) mini-photometer that is pre-programmed to perform a series of blood tests using pre-packaged reagent kits (Fig. 1). The Minilab is small (18 cm long × 10 cm wide × 4 cm high), lightweight (450 g) and portable (battery operated), making it ideal for field or hatchery situations where laboratory facilities are not available. To determine glucose, 5 µl of plasma and two drops of enzyme reagent are added to a test cuvette containing a pre-measured amount of reagent solution sufficient for one glucose determination. The contents are mixed thoroughly and incubated for about 15 min before analysis. The glucose reading is obtained by selecting the 'Gluc' function on the front panel, inserting the cuvette into the cuvette well and pressing the 'analyze' key. The digital display will show the result within 5 s. Detailed procedures for determining plasma glucose levels with the Minilab can be found in the Minilab operating manual supplied with the instrument.

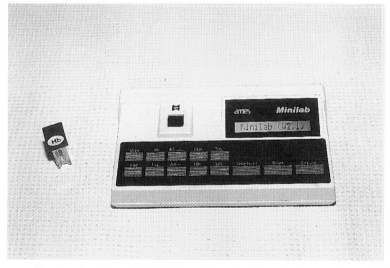

Fig. 1. The Ames Minilab for determination of glucose, hemoglobin and erythrocytes in blood.

Principle of the procedure and general comments

Each disposable cuvette is filled with 1 ml of reagent solution which consists of aminophenazone, hydroxybenzoic acid, 1,4-piperazine-diethanesulfonic acid (PIPES) buffer, detergent and stabilizers. The enzyme reagent contains glucose oxidase, peroxidase and mutarotase. Glucose oxidase in the enzyme reagent catalyzes the oxidation of glucose in the plasma. The resulting hydrogen peroxidase reacts with the aminophenazone and phenol, under the catalytic effect of the peroxidase, to form a colored complex. The intensity of the pink color produced is proportional to the amount of glucose originally present. The optical absorbance of the sample is measured at a wavelength of 546 nm and the absorbance value is converted to a plasma glucose concentration against an internal standard.

The Minilab values for glucose are very similar to those determined by a laboratory-based assay (Sigma Chemical Co., St. Louis, Mo., USA: diagnostic kit 315), differing by only 5% on average (Iwama *et al.*, 1995). The procedure is simple and the instrument is more portable and less expensive than a standard spectrophotometer. The main drawback for remote field use is that the procedure requires plasma, which must be separated from whole blood by centrifugation, and thus necessitates access to electric power. Considerations for sample handling and storage will be addressed in a later section.

ExacTech Blood Glucose Meter

The ExacTech Blood Glucose Meter (Medisense Inc., Cambridge, Mass., USA) is a portable electronic glucometer used by human diabetic patients (Fig. 2). The ExacTech Meter is a pen-sized instrument (14 cm in length, 1 cm in diameter) that measures blood glucose levels using ExacTech Blood Glucose Test Strips. The test strips are inserted by placing the end with silver contact bars into the end of the meter. A drop (20–30 μl) of whole blood is added to the target area of the strip and the test is begun by immediately depressing the control button at the opposite end of the pen. The display window will count down from 30 s and the blood glucose level will be displayed for 30 s before turning off. The company that manufactures the ExacTech Glucose Meter (MediSense, Inc.) has recently introduced the Pen 2 and the Companion 2 sensors which start automatically and decrease the testing time to 20 s. The ExacTech meters must be calibrated for each box of 50 test strips. The advantage of these glucometers over other models is that they do not require cleaning, wiping, blotting or timing; procedures which may introduce measurement error. The ExacTech is

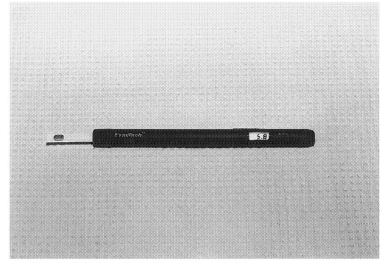

Fig. 2. The ExacTech Blood Glucose Meter.

faster and cheaper than the Minilab and does not require separation of plasma; however, there are limitations with the ExacTech pen not found with the Minilab. The ExacTech uses whole blood and extremes of hematocrit may therefore affect results; hematocrits over 55% yield lower values, hematocrits less than 35% yield higher values. The lower detection limit of the ExacTech pen is 2.2 mM (39.6 mg dl^{-1}) which may not be sufficient for resting or fasting fish. This compares to a lower limit of 1.1 mM (19.8 mg dl^{-1}) for the Minilab and Companion 2 sensor. The ExacTech Meter also tends to give slightly lower values than the Minilab, possibly due to the interference when using whole blood rather than plasma.

Principle of the procedure and general comments

When the blood sample is placed on the test strip, the glucose reacts with glucose oxidase on the strip to produce electrical microcurrents. The ExacTech Meter measures these microcurrents and converts them into a blood glucose value in mM or mg dl^{-1}. There are several models of portable glucometers available from retail pharmacy outlets, in addition to the one described here. Before being used in the field, these meters should be calibrated against a standard laboratory-based assay to confirm the validity of the readings.

Hemoglobin

Minilab

Hemoglobin measurements with the Minilab are based on the cyanmethemoglobin method, which is regarded as the most precise (Houston, 1990). Of whole blood, 5 μl is added to the test cuvette, mixed thoroughly and allowed to incubate for at least 1 min. The hemoglobin reading is obtained by pressing the 'Hb' function on the front panel, inserting the cuvette into the cuvette well and pressing the 'analyze' key. The digital display will show the result within 5 s.

Principle of the procedure and general comments

Each disposable cuvette is filled with 1.25 ml of reagent solution containing potassium ferricyanide, potassium cyanide, sodium chloride, phosphate buffer, and detergent. Hemoglobin is freed from red blood cells by lysis of the cell membrane. It is oxidized by the potassium ferricyanide–cyanide solution to form methemoglobin, which in turn is converted to the stable compound cyanmethemoglobin. The color intensity of the cyanmethemoglobin is directly proportional to the hemoglobin concentration in the sample and is measured photometrically at 546 nm to provide a quantitative determination of hemoglobin.

Hemoglobin concentrations determined with the Minilab show excellent (97%) agreement with a laboratory-based assay (Sigma Chemical Co., diagnostic kit 525-A) over a wide range of values (Iwama *et al.*, 1995).

Hemoglobinometer

The BMS Hemoglobinometer (Buffalo Medical Specialties Inc., Clearwater, Fla., USA) is a portable photometer for the evaluation of the hemoglobin content of the blood using the oxyhemoglobin method (Fig. 3). A drop of whole blood is placed upon a glass chamber and is hemolyzed with a wooden applicator coated with saponin (Fig. 4). When hemolysis is complete (15–25 s) the chamber is inserted into the compartment on the left side of the meter. When the instrument is brought up to the eye and the light button on the bottom is depressed, a split green field appears. The indicator button on the right side of the meter should then be moved until the field appears as one with no difference in intensity of color. The white line on the indicator button will line up with the corresponding hemoglobin concentration shown on the scale.

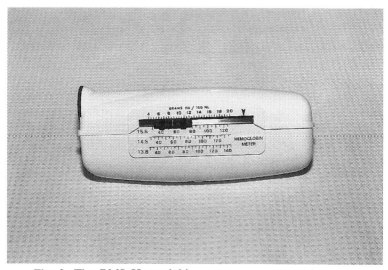

Fig. 3. The BMS Hemoglobinometer.

Fig. 4. Hemolysis, with a wooden applicator, of a blood sample for hemoglobin measurement with the BMS Hemoglobinometer.

Principle of the procedure and general comments

The BMS Hemoglobinometer compares the absorption of light through a defined layer of hemolyzed blood (oxyhemoglobin) to that of a standard glass wedge. An internal light source enters a diffusing glass window and passes through the hemolyzed blood sample and the standardized glass wedge simultaneously. A prism directs the light to a split-field ocular which contains a filter to adjust the transmission wavelength to 540 nm. The standardized glass wedge is moved by the indicator button until the light intensity is the same as that of the blood sample. The hemoglobin concentration is then read from the scale on the side of the instrument in g dl^{-1}.

The BMS Hemoglobinometer is able to closely match the laboratory-based assay in measuring hemoglobin values of 4 g dl^{-1} and above; however, levels less than 4 g dl^{-1} cannot be measured due to the detection limits of the meter (Iwama *et al.*, 1995). The Hemoglobinometer is cheaper than the Minilab system. However, it requires greater manual dexterity to operate and, because it depends on visual color discrimination, the readings can be rather subjective. It is recommended that if several people are using this instrument results from the same sample should be compared to assess operator sources of variation. Also, the 4 g dl^{-1} detection limit of the Hemoglobinometer may preclude quantitative measurements of hemoglobin in fish that are anemic.

Erythrocyte number

Minilab

The determination of erythrocyte (red blood cell) number with the Minilab is based on a turbidity measurement with Gower's solution. Of whole blood, 5 µl is added to the test cuvette using capillary tubes supplied with the reagent kit. The cuvette is shaken until the blood has mixed with the reagent solution and is allowed to incubate for at least 1 min. The erythrocyte reading is obtained by pressing the 'Ery' function on the front panel, inserting the cuvette into the cuvette well and pressing the 'analyze' key. The digital display will show the result within 5 s.

Principle of the procedure and general comments

Each disposable cuvette is filled with Gower's solution containing sodium sulfate and acetic acid. When the blood sample is added to the Gower's solution, the erythrocytes become spherical in shape, yielding a suspension the turbidity of which is measured photometrically at 546 nm. The turbidity of the sample is directly related to the number of erythrocytes per unit volume of blood and this conversion is performed by an algorithm in the Minilab.

Erythrocyte numbers measured with the Minilab are about three times higher than those determined by manual counts using a Neubauer hemacytometer (Iwama et al., 1995). The Minilab was designed for use with human blood and the erythrocyte procedure is based on turbidity measurements. Fish red blood cells are larger than human red blood cells and are nucleated. This may result in a denser solution that would overestimate the actual number of cells present and thus explain the large difference observed between the two methods. However, the relationship between the Minilab and manual method is linear over a wide range of values, and reasonable estimates of red blood cell numbers using the Minilab can be obtained by using the regression equation for that relationship (Fig. 5). Given that the manual method of counting red blood cells can be quite time consuming, the Minilab offers a rapid alternative for estimating erythrocyte numbers in fish.

Hematocrit and leukocrit

Hematocrit

Hematocrit (packed red blood cell volume; percent of total blood volume) determinations are commonly made with 70-μl microhematocrit capillary tubes, 75 mm long with 1.1–1.2 mm internal diameters, and coated with ammonium heparin to prevent clotting. Blood from the fish is drawn into the tube by capillary action and the end is closed with a plastic putty sealant (e.g., Critoseal). The blood cell and plasma fractions are separated by centrifugation in a standard microhematocrit centrifuge for 5 min at about 13,500 g. A microhematocrit capillary tube reader (e.g., Critocap) can then be used to measure the volume of packed red blood cells as a percentage of the whole blood volume in the tube. Hematocrit values are normally in the range of 30–50% depending on the size of the fish, and precisions of ±1% are normally attained with the microhematocrit tube reader. If a hematocrit tube reader is not available, simply measure the length of packed red cells and the length of red cells plus the plasma supernatant with a ruler. The hematocrit value is the ratio of these two measurements multiplied by 100. When sampling small fish where blood volume is limiting, it is often convenient to remove the plasma from the capillary tube after the hematocrit measurement and use the plasma for further analyses (e.g., glucose, ions, protein, etc.). Microhematocrit centrifuges require level surfaces and a source of electricity to operate, which can often

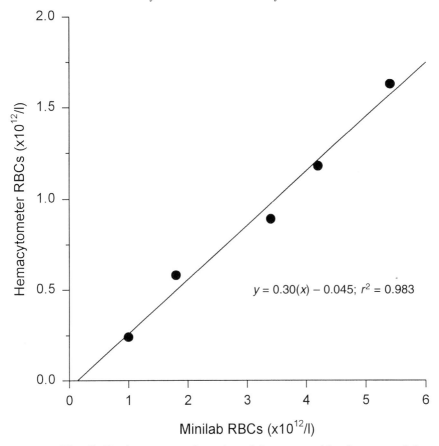

Fig. 5. Erythrocyte numbers in rainbow trout blood measured by manual counts and by the Minilab. Results of linear regression analyses are given in the equation, along with the r^2 value (adapted from Iwama, Morgan & Barton, 1995).

be found in a hatchery, but are less likely to be found in a more remote location such as a river system.

Leukocrit

Changes in the differential leukocyte count are one of the most sensitive indicators of acute stress in fish (Wedemeyer *et al.*, 1990). The leukocyte population is made up of lymphocytes, thrombocytes, neutrophils, and monocytes. Microscopic examination of blood smears stained with a colored fixative is the most accurate and precise method of determining white blood cell numbers. Typically, the number of lymphocytes per 1000 red blood cells is used as an indicator of leukocyte abundance. Detailed methods for counting blood cells can be found in Houston (1990) or Wedemeyer and Yasutake (1977). This technique, however, requires a fair amount of training, is quite time consuming, and requires rather expensive equipment (e.g., histological stains, compound micro-scope with 1000× magnification). The leukocrit is a more approximate, but rapid, measurement of white blood cell abundance and can provide an indication of fish health (McLeay & Gordon, 1977). Leukocrits are determined in a similar fashion to hematocrit values. After centrifug-ation of the blood sample in a microhematocrit tube, the thickness of the small white blood cell layer (buffy coat) separating the red blood cells from the plasma is measured with an ocular micrometer in a low-power (i.e., 50×) dissecting microscope. The leukocrit value (packed cell volume of leukocytes) is calculated as a percentage of the total volume in the tube as follows: (height of buffy layer/height of total blood volume) × 100. Leukocrit values in salmonid fish are normally in the range of 1–2% (Wedemeyer et al., 1990). If a dissecting microscope is not available, the leukocrit value may be estimated by the naked eye using the microhematocrit tube reader. In such cases, only crude estimates of the leukocrit value may be obtained (i.e., to the nearest 1%); however, this may be sufficient to detect significant increases in leukocytes (e.g., >5%) due to a subclinical infection (Wedemeyer et al., 1990).

Plasma constituents

Plasma protein

A protein refractometer is a convenient method for determining plasma protein concentrations. The ATAGO protein refractometer, for example, is a hand-held instrument (size: 17 cm length × 4 cm width × 4 cm height; 285 g) that can measure protein concentrations within a few seconds. One or two drops of plasma are placed on the glass prism and the cover plate is gently closed so that the sample spreads over the entire prism surface. The protein scale ($0–12$ g dl^{-1}) is then read through the eyepiece and the interception of the boundary line indicates the protein concentration, with a precision of $±0.2$ g dl^{-1}. Between measurements and after use, the glass prism surface should be cleaned with water and dried with lens paper to avoid scratching the surface. Also, the refractometer should be calibrated before use by placing a few drops of distilled water on the prism surface and adjusting the boundary line to the 'Wt' mark with the scale adjustment screw.

Principle of operation and general comments

The refractive index of a solution is related to its solute concentration. Proteins are the major constituents of plasma, therefore the refractive index can be used to measure protein concentrations. The ATAGO protein refractometer has been calibrated to use the refractive index method for measuring total protein in plasma or serum. The ATAGO protein refractometer shows good (88%) agreement with a laboratory-based protein assay (Sigma Chemical Co., diagnostic kit P-5656). The only disadvantage of the protein refractometer is that it requires about 40 µl of plasma to operate properly, compared to 10 µl for the clinical assay. This makes it somewhat impractical to use with small fish when sample volume is often limiting.

Plasma chloride

Effects of stress on osmoregulation can be determined by measuring plasma chloride ion levels with a Haake Buchler digital chloridometer. A total of 10 µl of plasma is added to 4 ml of acid reagent and the chloride ion concentration is automatically determined by coulometric titration using silver electrodes. A digital readout of the chloride concentration in mEq l^{-1} is obtained within 20 s. Detailed procedures can be found in the chloridometer instruction manual.

Although the initial equipment purchase price is quite high (approx. $5000 CDN or £2400), the chloridometer is easy to operate and samples can be processed rapidly at low cost. The instrument is well suited to a bench-top location in a hatchery facility, but needs alternating current and is therefore not portable enough for remote field use. It may be possible in a field situation, however, to freeze the plasma using a small container of dry ice and to store it for later analysis in the hatchery.

Physical condition indices

Length, weight, condition factor and organosomatic indices

Body length and weight are probably the most common and easily measured variables in fish biology and over time provide an important tertiary indicator of stress in fish culture, namely growth (Wedemeyer *et al.*, 1990). Decreased growth rates are often indicative of adverse environmental effects operating at lower levels of organization (i.e., cells, tissues, and organs). Lengths are generally determined using measuring boards or tape measures and the most common measures for fish are total, fork, and standard lengths (Anderson & Gutreuter, 1983).

Weighing fish in the field can be easily accomplished using inexpensive portable electronic balances (e.g., Ohaus portable balances). A more detailed discussion of techniques in the study of fish growth can be found in Busacker, Adelman and Goolish (1990).

The condition factor (K) is commonly used as an index of fitness or 'well-being' of fish, and is calculated from length (cm) and mass (g) measurements as:

$$K = \text{mass}/(\text{fork length})^3 \times 100 \qquad \text{(Ricker, 1975)}$$

As an indicator of stress, condition factors have been shown to decline in fish subjected to high rearing densities and other adverse environmental conditions (Goede & Barton, 1990). The use of condition factors in stress assessment studies should be interpreted with caution, however, as condition factors may also decline due to seasonal and developmental changes, such as the parr–smolt transformation (Vanstone & Markert, 1968).

Organosomatic indices can be calculated from field measurements made on selected tissues. The liver–somatic index (LSI) or hepatosomatic index (HSI) is the ratio of liver mass to total body mass expressed as a percentage [i.e., (liver mass × 100)/body mass]. The liver serves as a major storage site for glycogen and the HSI can therefore provide an index of the nutritional state of the fish (Anderson & Gutreuter, 1983; Busacker et al., 1990). The HSI may also be used as an indicator of growth potential in some fish populations (Adams & McLean, 1985). Other organ mass ratios which can be useful in stress studies include the viscerosomatic index (VSI), the gonadosomatic index (GSI) and the splenosomatic index (Goede & Barton, 1990). The assumption made with these indices is that lower than normal values indicate a diversion of energy away from organ growth in order to combat a stressor of some type. It is important to note that the various organosomatic indices may vary naturally with food availability, the state of sexual maturation and life history stage, often in concert with the season. These factors should be taken into account when attempting to use organosomatic indices in stress-related studies. Simple reproductive indicators which may also be used to assess stress effects on mature females include measuring fecundity (number of eggs per female) as well as egg diameter (Donaldson, 1990).

Autopsy-based condition assessment

The fish autopsy procedure was developed by Goede (1989; see also Goede & Barton 1990) to quantify the condition of fish based on external and internal examination of tissues and organs. Although not meant to be a diagnostic tool, the autopsy procedure can be used as a simple and rapid 'early warning system' to detect departures from a normal health–condition profile (HCP) before more sophisticated diagnostic techniques are employed. The desired sample size to establish an HCP is 20 fish. The fish should be examined soon after capture to prevent discoloration of the various organs. It is often convenient to lay the fish out in a row and begin with an assessment of the external features. Observations of fin damage, opercles, eyes, gills, pseudo-branchs, and the thymus gland are made according to the classification scheme and coding system described in Table 1. After the external observations are completed, the fish is opened up, using a pair of dissecting scissors, to expose the internal organs. A ventral cut from the anal vent forward to the pectoral girdle is usually the most efficient method to open the fish. The physical appearance of the liver, kidney, and spleen is evaluated, as well as the relative amount of mesenteric fat, hindgut inflammation and bile in the gall bladder (Table 1).

The autopsy data can be recorded on a standardized data sheet and summarized to provide the following information: (1) percentage of fish with normal and abnormal eyes, gills, pseudobranchs, thymus, spleen, hindgut, kidneys, and liver; (2) mean index values of damage to fins, thymus hemorrhage, mesenteric fat deposition, hindgut inflammation, and bile color. The calculations can be performed manually with the use of a pocket calculator. Data entry and calculations can also be performed in the field using a laptop computer and spreadsheet program. An alternative approach to the HCP is the calculation of a Health Assessment Index (HAI). In this system, all variables are assigned numerical values to calculate a single HAI and to allow statistical comparisons between samples (Table 1; Adams *et al.*, 1993).

Sampling considerations

Schedule of sampling

It is recommended that a routine monitoring program be established for a given fish population, to determine typical patterns in the meas-ured variables over time. We have found with cultured juvenile Atlantic salmon (*Salmo salar*), for example, that normal levels of glucose, hemoglobin, and erythrocytes can change during the first year of growth (Iwama *et al.*, 1995). It is important, therefore, to establish a pattern or trend in the variables of interest in order to detect meaningful increases or decreases due to a stressful event accurately. The frequency of sampling necessary to develop a physiological profile will depend

Table 1. *Description of variables and coding system used in the health condition profile (HCP) and health assessment index (HAI) fish autopsy systems (modified from Goede & Barton, 1990 and Adams et al., 1993). The letters used in the HCP column are defined in the Condition column and are generally, but not always, the first letter of the word. The numbers in the HAI column are relative indicators with no units. The variables are listed in the order in which they are usually examined, starting with the external features and moving to the internal organs.*

Variable	Condition	HCP value	HAI value
Fins	No active erosion	0	0
	Light active erosion	1	10
	Moderate active erosion	2	20
	Severe active erosion with hemorrhaging	3	30
Opercles	No shortening	0	Not
	Slight shortening	1	assigned
	Severe shortening, gills exposed	2	
Eyes	Normal	N	0
	Blind (one or both)	B	30
	Exophthalmic; swollen, protruding (one or both)	E	30
	Hemorrhaging (one or both)	H	30
	Missing one or both eyes	M	30
	Other	OT	30
Gills	Normal	N	0
	Frayed; erosion at tips of gills	F	30
	Clubbed; swelling at the end of gills	C	30
	Marginate; colorless margin along tips	M	30
	Pale in color	P	30
	Other	OT	30
Pseudobranchs	Normal; flat or concave in appearance	N	0
	Swollen and convex in aspect	S	30
	Lithic; white mineral deposits	L	30
	Swollen and lithic	S&L	30
	Inflamed; redness, hemorrhage	I	30
	Other	OT	30
Thymus	No hemorrhage	0	0
	Mild hemorrhage	1	10
	Moderate hemorrhage	2	20
	Severe hemorrhage	3	30

Table 1. *cont.*

Variable	Condition	HCP value	HAI value
Mesenteric fat	No fat deposits	0	Not assigned
	Less than 50% coverage of pyloric caeca with fat	1	
	50% of pyloric caeca covered with fat	2	
	More than 50% of caeca covered with fat	3	
	Pyloric caeca completely fat covered	4	
Spleen	Normal; black, very dark red, or red	B	0
	Normal; granular, rough appearance of spleen	G	0
	Nodular; cysts or nodules in the spleen	D	30
	Enlarged; noticeably enlarged	E	30
	Other	OT	30
Hindgut	Normal; no inflammation	0	0
	Slight inflammation or reddening	1	10
	Moderate inflammation or reddening	2	20
	Severe inflammation or reddening	3	30
Kidney	Normal; firm, red color, lying flat against the backbone	N	0
	Swollen or enlarged	S	30
	Mottled; gray discoloration	M	30
	Granular; granular appearance and texture	G	30
	Urolithiasis; creamy white deposits in the kidney	U	30
	Other	OT	30
Liver	Normal; red color	A	0
	Fatty or 'coffee with cream' color	C	30
	Nodules or cysts in the liver	D	30
	Focal discoloration	E	30
	General discoloration in whole liver	F	30
	Other	OT	30
Bile	Yellow-colored bile, gall bladder mostly empty	0	Not assigned
	Yellow bile, mostly full bladder	1	
	Light-green bile, full bladder	2	
	Dark-green to blue-green bile, full bladder	3	

upon the life cycle of the species as well as logistical considerations. In a culture situation, sampling on a monthly basis is generally adequate. The number of fish sampled will depend on the size of the population and the nature of the investigation. When working with cultured populations, obtaining sufficient numbers of fish is generally not a problem and 10–20 fish per treatment will suffice.

Evaluation of a stressful event such as grading and transport typically involves sampling fish before the procedure (control group), and at some point after the disturbance (treatment group). The magnitude of the responses observed will depend on the duration of the stressor and the time period between exposure to the stressor and blood sampling (Schreck, 1990). It is important, therefore, to understand the time courses of the response for the variables that are being measured. We have found in juvenile Atlantic salmon that blood glucose levels determined by the Minilab and the ExacTech Meter were significantly elevated 4 h after a grading procedure (Iwama et al., 1995). A similar time course for the glucose stress response using a laboratory-based assay was observed by Pickering, Pottinger and Christie (1982) during the recovery of brown trout (Salmo trutta) from acute handling. For most of the blood-related variables described in this chapter, 3–6 h after the stressful event is a reliable time interval to detect the various stress responses. Changes in gross features of organs are not likely to occur after a short period of disturbance, and the physical condition indices are therefore more suited to detect changes in fish condition over longer time periods (e.g., months).

Anesthetic

The fish should be handled carefully after capture and immobilized with an appropriate anesthetic before sampling. Tricaine methanesulfonate (MS-222) is the most commonly used fish anesthetic and is most effective at a concentration of 50–100 mg l^{-1} depending on fish size. Due to its acidic nature in soft freshwater, MS-222 solutions should be buffered with an equal weight of sodium bicarbonate ($NaHCO_3$) to prevent stressful effects of low pH (Summerfelt & Smith, 1990). Buffering with $NaHCO_3$ is not required in seawater, as it is sufficiently alkaline that MS-222 has little effect on pH.

Blood collection

The immobilized fish should be measured immediately for length in centimetres and mass in grams for growth rate and condition factor determinations. Blood can then be collected from the fish in a number of ways, depending on the size of the fish. For small fish (i.e., <20 g) sampling is terminal. The caudal peduncle is severed with a scalpel and blood is collected into heparinized microhematocrit tubes. With larger fish, a popular non-lethal method of sampling blood is to insert a syringe needle at the ventral midline just posterior to the anal fin and to continue the insertion at a 45° angle until it penetrates the caudal vessels lying between adjacent hemal arches (Fig. 6). This is most easily accomplished by inserting the needle until it stops against the backbone. The blood should then be drawn slowly into a heparinized plastic syringe. The total blood volume of salmonids is about 3–5% of body mass; therefore, care should be taken not to remove too much blood from the fish. As a general rule, removing 0.5% of total body mass (i.e., 0.5 ml of blood from a 100-g fish) will not affect survival.

Measurement protocol and sample storage

After blood collection, the fish can either be placed in a bucket of water to recover, or killed with a sharp blow to the head and set aside for the assessment of organosomatic indices and autopsy-based condition. The whole blood should then be used immediately to determine hemoglobin, erythrocyte and glucose values with the appropriate

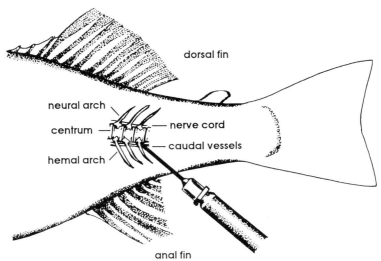

Fig. 6. Blood collection from the caudal vessels using a syringe (modified from Houston, 1990).

portable instrument. Red blood cells have been shown to swell when exposed to air (Strange, 1983), therefore hematocrits and leukocrits should also be determined as soon as possible after blood collection. The spun-down plasma fraction can then be removed from the capillary tube and used to measure glucose levels with the Minilab and protein concentrations with the refractometer. When using a syringe to collect blood from a large fish, excess whole blood may be expelled into a plastic 1.5-ml tube and spun down using a portable microcentrifuge (e.g., Millipore Personal Centrifuge) to separate the plasma from the cells. The plasma samples should be frozen in separate plastic tubes for later analysis of chloride ions. If time is a constraint, plasma glucose and protein concentrations may also be measured at a later date. A chest freezer capable of achieving a temperature of -20 °C is sufficient for storing plasma, although -80 °C is preferred for long-term sample storage. This procedure can also be accomplished in the field by using a portable generator for the centrifuge, and a cooler with dry ice to freeze the plasma until returning to the laboratory. If a source of power is not available for centrifugation, as in a remote field location, glucose and hemoglobin levels can still be determined on-site using whole blood with the ExacTech (or Companion) glucose meter and BMS Hemoglobinometer, respectively.

References

Adams, S.M. (ed.) (1990a). Biological indicators of stress in fish. *American Fisheries Society Symposium*, **8**, pp. 1–191.

Adams, S.M. (1990b). Status and use of biological indicators for evaluating the effects of stress on fish. *American Fisheries Society Symposium*, **8**, 1–8.

Adams, S.M. & McLean, R.B. (1985). Estimation of largemouth bass, *Micropterus salmoides* Lacépède, growth using the liver somatic index and physiological variables. *Journal of Fish Biology*, **26**, 111–26.

Adams, S.M., Brown, A.M. & Goede, R.W. (1993). A quantitative health assessment index for rapid evaluation of fish condition in the field. *Transactions of the American Fisheries Society*, **122**, 63–73.

Anderson, R.O. & Gutreuter, S.J. (1983). Length, weight, and associated structural indices. In *Fisheries Techniques*. Nielson, L.A. & Johnson, D.L. (eds.) pp. 283–300. American Fisheries Society, Bethesda, Maryland.

Busacker, G.P., Adelman, I.R & Goolish, E.M. (1990). Growth. In *Methods for Fish Biology*. Schreck, C.B. & Moyle, P.B. (eds.) pp. 363–87. American Fisheries Society, Bethesda, Maryland.

Cardwell, R.D. & Smith, L.S. (1971). Hematological manifestations of vibriosis upon juvenile chinook salmon. *Progressive Fish-Culturist*, **33**, 232–5.

Donaldson, E.M. (1990). Reproductive indices as measures of the effects of environmental stressors in fish. *American Fisheries Society Symposium*, **8**, 109–22.

Goede, R.W. (1989). *Fish Health/Condition Assessment Procedures.* pp. 1–31. Utah Division of Wildlife Resources, Fisheries Experiment Station, Logan, Utah.

Goede, R.W. & Barton, B.A. (1990). Organismic indices and an autopsy-based assessment as indicators of health and condition of fish. *American Fisheries Society Symposium*, **8**, 93–108.

Houston, A.H. (1990). Blood and circulation. In *Methods for Fish Biology*. Schreck, C.B. & Moyle, P.B. (eds.) pp. 273–334. American Fisheries Society, Bethesda, Maryland.

Iwama, G.K., Morgan, J.D. & Barton, B.A. (1995). Simple field methods for monitoring stress and general condition of fish. *Aquaculture Research*, **26**, 273–82.

McLeay, D.J. & Gordon, M.R. (1977). Leucocrit: a simple hematological technique for measuring acute stress in salmonid fish, including stressful concentrations of pulpmill effluent. *Journal of the Fisheries Research Board of Canada*, **34**, 2164–75.

Nikinmaa, M. (ed.) (1990). Vertebrate red blood cells: adaptations of function to respiratory requirements. *Zoophysiology*, **28**, 1–262.

Novotny, J.F. & Beeman, J.W. (1990). Use of a fish health condition profile in assessing the health and condition of juvenile chinook salmon. *Progressive Fish-Culturist*, **52**, 162–70.

Pickering, A.D., Pottinger, T.G. & Christie, P. (1982). Recovery of the brown trout, *Salmo trutta*, from acute handling stress: a time-course study. *Journal of Fish Biology*, **20**, 229–44.

Ricker, W.E. (1975). Computation and interpretation of biological statistics of fish populations. *Bulletin of the Fisheries Research Board of Canada*, **191**, 1–382.

Schreck, C.B. (1990). Physiological, behavioral, and performance indicators of stress. *American Fisheries Society Symposium*, **8**, 29–37.

Strange, R.J. (1983). Field examination of fish. In *Fisheries Techniques*. Nielson, L.A. & Johnson, D.L. (eds.) pp. 337–47. American Fisheries Society, Bethesda, Maryland.

Summerfelt, R.C. & Smith, L.S. (1990). Anesthesia, surgery, and related techniques. In *Methods for Fish Biology*. Schreck, C.B. & Moyle, P.B. (eds.), pp. 213–72. American Fisheries Society, Bethesda, Maryland.

Vanstone, W.E. & Markert, J.R. (1968). Some morphological and biochemical changes in coho salmon, *Oncorhynchus kisutch*, during parr–smolt transformation. *Journal of the Fisheries Research Board of Canada*, **25**, 2403–18.

Wedemeyer, G.A. & Yasutake, W.T. (1977). *Clinical Methods for the Assessment of the Effects of Environmental Stress on Fish Health.* pp. 1–18. U.S. Fish and Wildlife Service Technical Paper 89, U.S. Fish and Wildlife Service, Washington DC.

Wedemeyer, G.A., Barton, B.A. & McLeay, D.J. (1990). Stress and acclimation. In *Methods for Fish Biology*. Schreck, C.B. & Moyle, P.B. (eds.) pp. 451–89. American Fisheries Society, Bethesda, Maryland.

Fish species list

Common Name	Species Name
Adriatic sturgeon	*Acipenser naccarii*
African catfish	*Claria gariepinus*
Arctic char	*Salvelinus alpinus*
Atlantic cod	*Gadus morhua*
Atlantic salmon	*Salmo salar*
Blue gill	*Lepomis macrochirus*
Brook trout	*Salvelinus fontinalis*
Brown trout	*Salmo trutta*
Channel catfish	*Ictalurus punctatus*
Chinook salmon	*Oncorhynchus tshawytscha*
Chum salmon	*Oncorhynchus keta*
Coho salmon	*Oncorhynchus kisutch*
Common carp	*Cyprinus carpio*
Cutthroat trout	*Oncorhynchus clarki*
Dolphin fish (mahi mahi)	*Coryphaena hippurus*
European eel	*Anguilla anguilla*
Goldfish	*Carassius auratus*
Grass carp	*Ctenopharyngodon idellus*
Japanese parrotfish	*Oplegnathus fasciatus*
Lake trout	*Salvelinus namaycush*
Lamprey	*Petromyzon marinus*
Largemouth bass	*Micropterus salmoides*
Masu salmon	*Oncorhynchus masou*
Mozambique tilapia	*Oreochromis mossambicus*
Mummichog	*Fundulus heteroclitus*
Nile tilapia	*Oreochromis niloticus*
Northern pike	*Esox lucius*
Rainbow trout	*Oncorhynchus mykiss*
Red drum	*Sciaenops ocellatus*
Red gurnard	*Chelidonichthys kumu*
Sea bass	*Dicentrarchus labrax*
Snapper	*Pagrus auratus*

Spotted sea trout	*Cynoscion nebulosis*
Striped bass	*Morone saxatilis*
Turbot	*Scophthalmus maximus*
Walleye	*Stizotedion vitreum*
White sucker	*Catastomus commersoni*

Index